CULTURAL INTELLIGENCE FOR WINNING THE PEACE

CULTURAL INTELLIGENCE FOR WINNING THE PEACE

Edited by Juliana Geran Pilon

THE INSTITUTE OF
WORLD POLITICS PRESS
WASHINGTON, D.C.

THE INSTITUTE OF WORLD POLITICS PRESS
1521 16th Street NW
Washington, D.C. 20036

Cover photo: "Yard Side Talk – U.S. Army Staff Sgt. Fred Hampton kneels on a knee to talk with a young Iraqi boy at the future site of a park in the Thawra 1 neighborhood of the Sadr City district of Baghdad on June 20, 2008." Hampton is assigned to the 1st Battalion, 6th Infantry Regiment, Task Force Regulators. 6/27/2008 U.S. Air Force photo by Tech. Sgt. Cohen A. Young.
Courtesy www.defenselink.mil.

"Adda Bozeman and a Bygone Tradition in Foreign Affairs Analysis that Must Be Revived," by John Lenczowski, appeared originally in *The National Security Studies Quarterly*. Reprinted with permission.

"Culture Clash-ification: A Verse to Huntington's Curse" (chapter 3), by Frederick S. Tipson, appeared originally in *Foreign Affairs*. Reprinted with permission.

"Islamism and Stratagem" (chapter 16), by John J. Dziak, appeared in *The Intelligencer*. Reprinted with permission.

"War and the Clash of Ideas" (chapter 2), by Adda Bozeman, appeared originally in *ORBIS*. Reprinted with permission.

The views expressed in this anthology are those of the original author(s), and not necessarily those of The Institute of World Politics, the United States government, or any other person or institution.

Please visit us online at www.iwp.edu.

First edition

ISBN-13: 978-0-615-51939-5
ISBN-10: 0615519393

Contents

Preface
and
Acknowledgements

Currently it is widely accepted that understanding the challenges to national security is impossible without appreciation of the cultural dimension, sometimes referred to as "the human terrain." Yet when I first taught this course in the 1990s, I felt that I had to justify the intrusion of soft-discipline material into an otherwise hard-power syllabus. Until, suddenly, everything changed.

I will never forget that day. I routinely assigned Samuel Huntington's audacious and provocative *The Clash of Civilizations* for the initial session, to get the discussion going and outline the major relevant concepts: culture, civilization, and conflict. Of course, I don't usually expect students to have read a whole book prior to the first day of classes. But that particular year, I wasn't even sure anyone would show up. The course had been scheduled to start on September 12. The year was 2001.

I did not really expect to teach, half-expecting school to be cancelled. I left the house mainly to get away from the nonstop news, half-expecting an empty classroom. To my amazement, I found a standing-room only audience: far more people than had been listed on my class roster sheet. Yet no one said a word. Even more astonishing, it turned out this was not because they had failed to do the reading: they had done it, all right. The crowd was there because these students wanted to understand what had just happened, and were speechless with disbelief. Our civilization had been dealt a huge blow. Culture clearly mattered; but how, and why?

From that day on, the course needed little justification. Each semester a new group of students added their perspectives, helping me to see new angles and approaches to this complex topic. Given the constantly changing security environment and the growing research in this area, I never taught it quite the same way twice: each time, I would add another book or article by experienced and brilliant practitioners who shared their invaluable experiences in dangerous and complex new theaters like Iraq and Afghanistan. Yet the basic themes have stayed the same, which is why I thought it useful to compile some of the best writings in one anthology. It goes without saying that I am enormously grateful to all my excellent colleagues and the fine students who have helped me shape its present format; this book could not have been produced without them.

First of all, I want to thank John Lenczowski, President of The Institute of World Politics, for introducing me to the magnificent Adda Bozeman, a historian and strategic thinker who deserves to be far better known: for one thing, it was she, rather than Huntington, who initiated the "clash of civilizations" approach to international relations. I am honored to include her in this anthology.

I also want to thank my colleagues Brigadier General Walter Jajko, and Professors J. Michael Waller, Marek Chodakiewicz, Jack Tierney, and Ken deGraffenreid for the ideas they have shared with me on the topics specifically addressed in this anthology; and to Jack Dziak, Montgomery McFate, and Andrew Garfield, authors included in this book, for their rich experience and advice. LTC Carl G. (Glenn) Ayers of SAIC (Science Applications International Corporation), until recently head of psychological operations at the Defense Department, has offered not only many fascinating insights into these operations but also shared many useful additional studies. Last but hardly least, I am grateful to the capable and diligent Charles Van Someren, whose work on the logistical aspects of this book finally made possible its release.

Today, the vital importance of culture for understanding statecraft is taken as a given throughout the intelligence and defense communities. But it will take a while before a new approach to national security education takes hold in academia. I hope this anthology will play at least a modest role in the painstaking effort to recalibrate the graduate curriculum to more adequately meet the daunting challenge of promoting a safe peace in the 21st century. America is a peaceful nation. We don't want to wage war; we wish to wage peace — yet the latter, in many ways, is harder to win. In order to not have to fight, we must be smarter and better prepared.

List of
Contributors

BG Ralph O. Baker, U.S. Army, is the assistant deputy director of the Po-
litico-Military Affairs, J5, Strategic Plans and Policy Directorate-Middle East,
under the Joint Chiefs of Staff, at the Pentagon. He holds a B.S. from the
United States Military Academy, an M.A. from Central Michigan University,
and an M.S. from the Naval War College. BG Baker has served in a variety
of command and staff positions in the continental United States, the United
Kingdom, Germany, and Iraq.

Adda Bozeman was professor emeritus of history and international relations
at Sarah Lawrence College, where she taught from 1947 to 1977. She gradu-
ated from l'Ecole Libre des Sciences Politique in Paris, became a lawyer at the
Middle Temple Inn of Court in London, and practiced with an international
firm in Berlin, The Hague, and London before moving to the United States in
the late 1930s. Her influential scholarly books include *Politics and Culture in
International History: From the Ancient Near East to the Opening of the Modern Age*,
Strategic Intelligence & Statecraft: Selected Essays, *Conflict in Africa: Concepts & Realities* and *Regional Conflicts Around Geneva*. Shortly before her death in 1994, she
lectured at The Institute of World Politics.

John J. Dziak leads the work of IASC (the International Assessment and
Strategy Center) on technology security, strategic denial and deception and
countermeasures. He has served over three decades as a senior intelligence
officer and executive in the Office of the Secretary of Defense and in the
Defense Intelligence Agency, with long experience in weapons proliferation
intelligence, counterintelligence, strategic intelligence, global countermeasures and intelligence education. In addition to IWP, he has also taught at

George Washington University and Georgetown University. The author of *Chekisty: A History of the KGB*, *The Military Relationship Between China and Russia*, *Soviet Deception: The Organizational and Operational Tradition*, and *Themes of Soviet Strategic Deception and Disinformation*, he is also co-author of *Bibliography on Soviet Intelligence and Security Services*.

Antulio J. Echevarria II serves as Director of Research for the U.S. Army War College, following a military career of 23 years, having held a variety of command and staff assignments in Europe and the United States. Dr. Echevarria is a graduate of the U.S. Military Academy, the U.S. Army Command and General Staff College, and the U.S. Army War College, and holds M.A. and Ph.D. degrees in history from Princeton University.

LTC Lester W. Grau, U.S. Army, Retired, is a military analyst in FMSO. He has a B.A. from the University of Texas at El Paso and an M.A. from Kent State University, and is a graduate of the Command and General Staff College, the U.S. Army Russian Institute, the Defense Language Institute, and the U.S. Air Force War College. He has served in various command and staff positions in CONUS, Europe, and Vietnam.

Col. T. X. Hammes, U.S. Marine Corps, Retired, served thirty years in the Marine Corps at all levels in the operating forces, to include command of a rifle company, weapons company, intelligence company, infantry battalion, and the Chemical Biological Incident Response Force. His final tour in the Marine Corps was as Senior Military Fellow at the Institute for National Strategic Studies, National Defense University. He is the author of *The Sling and the Stone: On Warfare in the 21st Century*.

Sheila Miyoshi Jager is an Associate Professor of East Asian Studies at Oberlin College. She is currently a Visiting Research Professor of National Security Studies for the Strategic Studies Institute at the U.S. Army War College. Her publications include *A Genealogy of Patriotism: Narratives of Nation-building in Korea* and, with Rana Mitter, *Ruptured Histories: War, Memory and the Post-Cold War in Asia*. Her book, *Korea: War Without End*, is forthcoming.

John W. Jandora is the supervisory threat analyst at U.S. Army Special Operations Command and the author of two books and numerous articles on warfare and Middle Eastern history.

COL Gregory Julian, U.S. Army, is currently in the Office of the Secretary of Defense OSD Public Affairs, Joint Communications.

Jacob Kipp, U.S. Army, is the Director of the Foreign Military Studies Office (FMSO), Fort Leavenworth, Kansas. He has a B.S. from Shippensburg University and an M.A. and Ph.D. from Pennsylvania State University.

John Lenczowski is the Founder and President of The Institute of World Politics, an independent graduate school of national security and international affairs in Washington, D.C. His M.A. and Ph.D. are from Johns Hopkins University's School of Advanced International Studies. He has served as Special Assistant to the Assistant Secretary of State for European Affairs, Special Adviser to the Under Secretary of State for Political Affairs, consultant to the Human Rights and Humanitarian Affairs Bureau at the Department of State, and Director of European and Soviet Affairs at the National Security Council. In the last of these capacities, he served as White House principal adviser on Soviet affairs. He is the author of *Soviet Perceptions of U.S. Foreign Policy* and *The Sources of Soviet Perestroika*.

Montgomery McFate is a cultural anthropologist with the Joint Advanced Warfighting Program at the Institute for Defense Analyses. She has also served as an American Academy for the Advancement of Science fellow at the U.S. Navy's Office of Naval Research (ONR), where she was awarded a Distinguished Public Service Award by the Secretary of the Navy.

COL John J. McCuen, U.S. Army Retired, is an author and consultant on counterinsurgency warfare. A 1948 graduate of the U.S. Military Academy, he holds an M.A. in international affairs from Columbia University. Throughout his long and distinguished military career, COL McCuen served in a variety of command and staff positions in the United States, Vietnam, Thailand, Germany, and Indonesia, where he was chief of the U.S. military assistance group (U.S. Defense Liaison Group, Indonesia). In 1966 he published *The Art of Counter-Revolutionary War—The Strategy of Counter-Insurgency*, a prescient and seminal work on irregular warfare, listed as a "counterinsurgency classic" in the U.S. Army/Marine Corps Counterinsurgency Field Manual (FM 3-24).

Juliana Geran Pilon, Ph.D. (University of Chicago) is a leading scholar of the cultural aspects of strategy, international affairs, democratization, and national security. She teaches at The Institute of World Politics in Washington, D.C. and the National Defense University. As Vice President for Programs at the International Foundation for Election Systems (IFES) during the 1990s, she designed, conducted, and managed projects related to a wide variety of democratization projects. Her previous books include *Why America Is Such a Hard Sell: Beyond Pride and Prejudice*; *The Bloody Flag: Post-Communist Nationalism in East-Central Europe*; *Every Vote Counts: The Role of Elections in Building Democracy*; and *Notes from the Other Side of Night*.

LTC Karl Prinslow, U.S. Army, Retired, manages the Joint Reserve Intelligence Center (JRIC) at Fort Leavenworth, Kansas. He has a B.S. from the U.S. Military Academy, an M.A. from the Naval Postgraduate School, and an MBA from Baker University. A former foreign area officer (Africa), LTC Prinslow also served in a variety of command and staff positions in the infantry.

LTC Fred Renzi, U.S. Army, is a psychological operations officer. He holds degrees from the Naval Postgraduate School, the United States Military Academy at West Point and is a graduate of the U.S. Army Command and General Staff College. LTC Renzi has held various command and staff positions in Europe and the continental United States. He deployed with the 1st Armored Division to Operations Desert Shield and Desert Storm and with the 1st Psychological Operations Battalion (Airborne) to Haiti.

CPT Don Smith III, U.S. Army Reserve, is an action officer at the Foreign Military Studies Office, Ft. Leavenworth, Kansas.

Lt. Col. George W. Smith, U.S. Marine Corps, shared a first place award with this essay, written while attending the Marine Corps War College. He served previously as plans officer, I Marine Expeditionary Force, and is currently commanding officer, 1st Force Reconnaissance Company, Camp Pendleton, California.

Fred Tipson is at AT&T, though not the Verse Division. The views expressed (you might have guessed) are lacking supervision.

MAJ Christopher H. Varhola, U.S. Army Reserve, is a cultural anthropologist whose current research focuses on religious identity and practice along the Swahili Coast in Tanzania. He holds a B.S. from The Citadel, an M.S. in International Relations from Troy State University, and a Ph.D. from Catholic University. Previous assignments include cultural affairs officer with the 352d Civil Affairs Command in Iraq, researcher for the Center for Army Lessons Learned in Iraq, cease-fire monitor in the Sudan, and a variety of command

and staff positions in the United States, Saudi Arabia, and Africa.

LTC Laura R. Varhola, U.S. Army, is the Defense and Army Attaché, U.S. Embassy, Dar es Salaam, Tanzania. She received a B.A. from the University of Michigan, an M.S. in International Relations from Troy State University, and is a graduate of the French Command and General Staff College. She has served in a variety of command and staff positions in the United States, Korea, Latin America, Iraq, and Africa.

Debra D. Zedalis was the Chief of Staff for the Installation Management Agency, Europe Region, Heidelberg, Germany. She has worked for the U.S. Army in Europe since 1988, serving as the Chief of the Management Division, Office of the Deputy Chief of Staff for Resource Management, as well as the Chief of the Installation Management Support Division, Office of the Deputy Chief of Staff for Personnel and Installation Management.

1

How Cultural Intelligence Matters

Juliana Geran Pilon

That is why it was called Babel, because there the Lord confounded the speech of the whole earth; and from there the Lord scattered them over the face of the whole earth.
—Genesis 11: 9

Our ancestors understood and accepted the limitations of our nature and our mortal predicament. Yet they found comfort in the cyclical wonder of Nature's perennial blissful regeneration, thus offering unmistakable proof – were any needed – that in the midst of inevitable change, decay, and death, there is rebirth and joy. The Sacred was the revolving predictability of order as against the accidental, i.e., meaningless, Profane, which interrupted the harmonious Cycles with dissonance.[1] Out of strife comes harmony, out of despair, hope. Each day brings a new reminder that night is followed by day, the Planets revolving as perfect Spirits guided by a benevolent Power beyond comprehension. The seeming dialectical opposition of death and life is in fact a continuum, as each end turns in some ways into a new beginning, the ensuing synthesis a link in the chain of Being whose defining quality is transformation, movement. While the Greek *agon* is the root of "agony," the word simply refers to competition, which is central to all existence and forms the basis of evolution.

To be is to be transformed. Inevitably, life is a series of battles interrupted by moments of peace, whose ephemeral beauty is cherished all the more for its brevity. The lot of our species is transformation - like Heraclitus's river,[2] where nothing is ever the same even as the wheel turns and turns.

---

1   Mircea Eliade, *The Sacred and the Profane and Cosmos and History: The Myth of Eternal Return* (New York: Harper & Row, 1969), esp. Chapter 3: "Misfortunes and History."

2   G.S. Kirk & J.E. Raven, *The Pre-Socratic Philosophers: A Critical History with a Selection of Texts,*

By contrast, his fellow pre-Socratic cosmologist Parmenides thought that beyond everything was the One Reality.[3] One can picture that monist vision naturally, almost seamlessly, morphing into the monotheism of the world's major civilizations. Yet this monolithic, divinely sanctioned self-righteousness enhanced the intolerance that led to violence beyond the scope of earlier, more pluralist polytheistic societies.

Our Biblical ancestors considered the multiplicity of languages, cultures, and rituals to be a curse, bound to result in discord and misunderstanding, as Divine punishment for our arrogance, selfishness and obtuseness. Having now arrived at a stage in history when we can easily destroy our entire species in even less time than it took the God of Genesis to fashion us in His image, the consequences of conflict can be cataclysmic. While hardly less sinful than his primordial ancestors, modern man might at least dare to imagine a world in which there is greater understanding, compassion, and – is there even such a concept? – peace. At our most cynical moments, we despair; then recoil to hope – perhaps against hope – that some measure of human understanding can yet be brought to this little piece of the universe that we were so generously allotted.

This does not have to mean indulging in a utopian dream of everlasting bliss, where everyone is tolerant of everyone else, although so deeply is this postulate entrenched in the Western, and in particular American, *Weltanschauung* that little can be done to prevent it. Surely some day we will all get along, all disagreements will be dialectical, and reason will prevail. The late historian Adda Bozeman exposed this culturally engrained penchant for wishful thinking:

> [t]he premise is widely accepted today that clashes of ideas are somehow either irrational departures from the ground rules of normal behavior or ruses to cover up peace-defying policies…. Nowhere outside North America and Northern Europe does one encounter the overriding desire to avoid armed conflict and to seek peaceful settlement of disputes that leading peace-minded scholars in our society assume to generally present.[4]

The humanist impulse, the catechism of scientific progress, and the constitutionally-enshrined pursuit of happiness in the here-and-now, have all conspired to transform the hope for peace on earth from a vain illusion into

---

(Cambridge, UK: Cambridge University Press, 1969), ch. VI "Heraclitus of Ephesus."

3   Ibid., ch. X, "Parmenides of Elea."

4   Adda B. Bozeman, "War and the Clash of Ideas," *Orbis*, Spring 1976, pp. 64-65.

a virtual entitlement. The idea that strife and respite are merely alternating moments along the trajectory of history is anathema to the Western – especially the modern Western – mind. As a result, our entire approach to strategy and intelligence-gathering has been flawed. In her superb collection *Strategic Intelligence & Statecraft: Selected Essays*,[5] Bozeman describes the cultural gulf that separates us from civilizations with a diametrically opposed approach to conflict and war, exposing the misconceptions that undermine our strategic intelligence.

The founder and president of The Institute of World Politics, Professor John Lenczowski, summarized Professor Bozeman's contribution to the study of international affairs after her death in 1994:

> There were several features that characterized Prof. Bozeman's work that made it distinctive enough to be called a 'tradition.' The first was its emphasis on the study and appreciation of history – not just recent diplomatic history, but the sweep of the history of Western and other major civilizations – and its relevance to contemporary foreign policy. The second was its emphasis on taking ideas, values, religion and other belief systems seriously. The third was its focus on analyzing foreign political cultures on their own terms and avoiding the pitfalls of "mirror-imaging" – a practice of assuming that foreigners think and act "just like us." ... The U.S. intelligence community, for example, has consistently disregarded the importance of having its analysts read the character-revealing literature of foreign cultures.[6]

Such questions do not often bother the average knowledge-challenged American going about his daily business simply wishing everyone spoke English and stocks were up again. We don't much like to be bothered with International Studies, let alone Philosophy of History. But one thing our co-nationals have demonstrated is that when attacked, they retaliate. It takes a while for this mammoth nation to awaken, but when it does, it thunders with rage. So it was after September 11. The murders in New York, Washington, and Pennsylvania, stunned the nation. Unable to fathom how such an outrage would happen in plain daylight on their own soil, Americans were truly mystified: how could these murderous zealots, bent on imposing their one language, one mindset, one book, over the entire earth, kill themselves and thousands of innocent men, women, and children?

Americans went to war in Afghanistan because they felt that war had

---

5   Adda B. Bozeman, *Strategic Intelligence & Statecraft: Selected Essays* (Brassey's Intelligence and National Security Library, 1992).

6   John Lenczowski, "Adda Bozeman and a Bygone Tradition in Foreign Affairs Analysis that Must Be Revived," *National Strategic Studies Quarterly*, 1995.

been declared against the U.S. – against the American people. The nation was mobilized: we would take our ammunition and blast the vicious enemy into oblivion so he would never dare attack us again. Seeking Bin Laden and his murderous Al Qaeda, American soldiers drove out the barbaric Taliban who had given him shelter, then turned to Iraq, where Saddam had been taunting America and the world for over a decade – and indeed had been found to be hoarding chemical weapons in 1992. President Bill Clinton had been preparing to strike Iraq for years; but it was George W. Bush who carried through and, in a manner, won the ground war. Saddam was caught, tried, and executed; his country was rid of its tyrant, and the U.S. of a dangerous enemy.

And then it hit us: we could not declare full victory – and not just because Bin Laden was still at large. We had forgotten that we still had to win the peace – having (wrongly) assumed that it would come as soon as the Iraqi population was free to hold elections.[7] In order to leave behind a stable nation, we had to be not only stronger but smarter. We had to understand our enemies as well as our friends; we even had to recognize that the two are not quite mutually exclusive.

In fact, we would first have to understand ourselves. It dawned on us that we were sometimes our worst enemies. We had come face to face with the quagmire of "hybrid" warfare which Colonel John McCuen defines as

> [a] combination of symmetric and asymmetric war in which intervening forces conduct traditional military operations against enemy military forces and targets while they must simultaneously—and more decisively—attempt to achieve control of the combat zone's indigenous populations by securing and stabilizing them (stability operations). Hybrid conflicts therefore are full spectrum wars with both physical and conceptual dimensions: the former, a struggle against an armed enemy and the latter, a wider struggle for, control and support of the combat zone's indigenous population, the support of the home fronts of the intervening nations, and the support of the international community. In hybrid war, achieving strategic objectives requires success in all of these diverse conventional and asymmetric battlegrounds.[8]

No easy task; and it doesn't help that most Americans haven't a clue about other cultures. What is worse, foreigners don't understand ours either – but don't even realize it. Far from improving, a few millennia after Babel, global

---

7   For an excellent and detailed account of CIA assessments and predictions relating to the Iraq war, see *COBRA II: The Inside Story of the Invasion and Occupation of Iraq*, by Michael R. Gordon and General Bernard E. Trainor (New York: Random House, 2007).

8   COL John J. McCuen, USA Ret., "Hybrid Wars," *Military Review*, March-April 2008, www.usacac.army.mil/CAC/milreview/English/MarApr08/McCuenEngMarApr08.pdf

understanding had greatly deteriorated.

The post-9/11 reality hit us hard: as a rule, we tend not to speak any foreign languages, and often have trouble with our own. In addition, it is easy to commit the most insidious of intelligence fallacies: mirror-imaging, which refers to the propensity to think that others are just like us, acting from similar motives and making similar rationality calculations. A seemingly opposite yet closely related malady is what psychologists call "fundamental attribution error." Put simply, it is the assumption that people with whom we are not familiar are fundamentally different from us in morally sinister ways. Either way, post-Babel reality leads to lethal cacophony. Different human communities are similar as well as different in unexpected ways that may be learned, but it all takes time.

Fortunately we are slowly beginning to realize, at the highest levels in the strategic and intelligence community, and even in the U.S. Congress, that the task at hand is complex: we have to tackle and master nothing less than other mindsets, other cultures. The post-Babel world turned out to be far more inscrutable and immeasurably more dangerous than it seemed at first blush.

Culture and intelligence

The term "culture" is relatively new, even if the general idea permeates throughout the great classic writings of antiquity, including Homer, Herodotus, and Chinese scholars of the Han dynasty,[9] all of whom recognized the existence of different habits of thinking and behavior. The generic Latin retains the primary notion of "cultivation," with its implication of purpose, design, and form. Its first appearance during the nineteenth century in Germany as *Kultur*, while reflecting an interest in the variety of human expression, already contained the normative evaluation that would come to the fore later, with the assumption that some cultures were more "civilized" than others. Curiously, the concept of culture took longer to penetrate the public consciousness in France and England. It did, however, resonate in the U.S., undoubtedly as a result of the complex ethnicity of our population. No less important was the new speed of communication at the turn of the 20th century, followed by that misguided conflagration which, far from being the War to End all Wars, inaugurated a period of unprecedented slaughter. And

---

9   A. L. Kroeber and Clyde Kluckhohn, *Culture: A Critical Review of Concepts and Definitions* (New York: Vintage Books, Random House, 1952), Introduction and Part I, "General History of the Word Culture."

while the commonplace that its sequel was merely Great War Two is a bit simplistic, there is little doubt that both conflicts were greatly exacerbated by the cultural indifference, at times even illiteracy, exhibited by the leadership of the Great Powers in 1919.[10]

The anthropological fashion that took hold in America, which crystallized around the idea of custom, as well as the emerging fields of sociology and behavioral psychology, soon led to a virtual explosion of methodologies and theories in what came to be known rather loftily as "the social sciences." Definitions of culture proliferated. Variously classified as either descriptive (emphasizing the empirical dimensions), historical (which required more extensive field studies, including archeology), normative (based on philosophical, religious, and political values), psychological, structural (involving analyses of the social stratification), and genetic (which may or may not have been connected with the psychological analyses), by the middle of the century over one hundred and sixty definitions could be identified in the United States alone.

That said, the concept of culture common to all scientific approaches involves a set of attributes and products of human societies, transmitted by means other than biological heredity, which are essentially lacking in sub-human species but are characteristic of human beings living in social groups.[11] Cultures differ from one society to another, yet they interact and evolve even as they retain elements that persist through time, for reasons as varied and complex as humanity itself. Their variety is a source of both wonder and grief. Whether we embrace it or ignore it, we cannot will it away.

We certainly tried. In the euphoria that followed incredulous shock after the astonishingly rapid demise of the Soviet Bloc, many in the West declared "an end of history" as culture was expected to become obsolete. Not by everyone, of course. In addition to Adda Bozeman, there was also Princeton historian Bernard Lewis. In lucid, elegant prose he specifically warned against Islam: "It should now be clear that we are facing a mood and a movement for transcending the level of issues and policies and the governments that pursue them. This is no less than a clash of civilizations – that perhaps irrational but surely historic reaction of an ancient rival against our Judeo-Christian heritage, our secular present, and the worldwide expansion of both. It is crucially important that we on our side should not be provoked into an equally historic

---

10   For a superb detailed account, see Margaret Macmillan, *Paris 1919: Six Months that Changed the World* (New York: Random House, 2002).

11   Kroeber and Kluckhohn, op. cit., p. 284.

but also equally irrational reaction against that rival."[12]

Unfortunately, the late Samuel P. Huntington, who appropriated the expression for use in the seminal text *The Clash of Civilizations and the Remaking of World Order*, did not heed that warning. After declaring, with inexplicable certitude, that "the underlying problem for the West is not Islamic fundamentalism," he accused the religion itself: "it is Islam, a different civilization whose people are convinced of the superiority of their culture and are obsessed with the inferiority of their power."[13] It is Islam itself, not the extreme views of Salafist, Wahabist, and other radical terrorists, that fuels the conflict "between Islam and the West," according to Huntington.

After somewhat casually dividing the human race into "seven or eight" civilizations (defined as cultures writ large[14]) differentiated primarily in terms of religion,[15] he asserts that "people who share ethnicity and language but differ in religion may slaughter each other"[16] and indeed do. The central theme of his book is that civilizations "are shaping the patterns of cohesion, disintegration, and conflict in the post-Cold War world."[17] But then how are we to explain the countless examples of cooperation among different religious groups throughout the world? Conversely, internecine warfare has characterized virtually every religion at some point, and continues today – notably among Sunni and Shia, to say nothing of the slaughter going on inside these sects themselves.

Especially puzzling is his approach to the following question: "What about Jewish civilization? Most scholars of civilization hardly mention it. In terms of numbers of people Judaism clearly is not a major civilization"[18] – nor, it would seem, even a minor one. Even more astonishing, this point is made... in a footnote. How, then, to explain the conflict between Muslims and Jews? For that is surely no minor turmoil (unlike, say, an internecine skirmish –

---

12 Bernard Lewis, "The Roots of Muslim Rage: Why So Many Muslims Deeply Resent the West and Why Their Bitterness Will Not Be Easily Mollified," *Atlantic Monthly*, 266 (September 1990), p. 60. Emphasis added.

13 Samuel P. Huntington, *The Clash of Civilizations and the Remaking of World Order* (New York: Simon & Schuster, 1996), p. 217.

14 "A civilization is the broadest cultural entity." Ibid., p. 43.

15 "To a very large degree, the major civilizations in human history have been closely identified with the world's great religions." Ibid., p. 42.

16 Ibid.

17 Ibid., p. 20.

18 Ibid., p. 48.

however genocidal in scope – inside the continent of Africa).[19]

Huntington does not make it easy on himself. Unlike Lewis, for example, he does not adopt the term "Judeo-Christian" to describe liberal values. "Western" civilization is distinguished from the "Orthodox" (which is after all still Christian) and the "Latin American" (predominantly Catholic). Charging that "the West's sponsorship, at the height of its power vis-à-vis Islam, of a Jewish homeland in the Middle East laid the basis for ongoing Arab-Israeli antagonism"[20] might suggest that he considers Israel to be part of the Western sphere. Not so: Huntington repeatedly lists Israel as a non-Western country. How can a theory that claims to attribute all major post-Cold War conflicts to rifts along civilizational [sic] "fault lines" while failing to place Israel within a civilization expect to get away with it? The disclaimer that Huntington includes in his preface, warning his readers that "this book is not intended to be a work of social science,"[21] would be amusing were it not perfectly serious.

One is hard pressed to find a more appropriate rebuttal than Frederick S. Tipson's clever verses:

> His model may appear to be well-grounded in the past,
> But Huntington has pitched his product much too hard and fast.
> As he explores these culture wars, he seems to be inviting
> What used to be a weakness of the school of "realist" writing,
> Ascribing to a concept, like a culture or a state,
> A physical reality which doesn't quite equate.[22]

Levity aside, there is no denying that Huntington's influence on the study of international affairs has had many positive aspects. He is rightly to be credited with having re-established culture as a legitimate subject for study by political scientists, who had heretofore been virtually oblivious to it.[23] Admittedly, it took a while. A year after the publication of *The Clash*

---

19 "The bloody clashes of tribes in Rwanda has consequences for Uganda, Zaire, and Burundi, but not much further." Ibid., p. 28. Elsewhere, he notes that anyway "most major scholars of civilization except Braudel do not recognize a distinct African civilization." P. 47.

20   Ibid., p. 263.

21   Ibid., p. 13.

22   Frederick S. Tipson, "Culture Clash-ification: A Verse to Huntington's Curse," *Foreign Affairs*, March/April 1997, www.foreignaffairs.org/.../frederick-s-tipson/culture-clash-ification-a-verse-to-huntington-s-curse.html

23   In a study for the Rockefeller Brothers Fund, Amir Pasic references Huntington's *Clash of Civilizations* in order to illustrate the statement that "ever since the end of the Cold War, culture has made a dramatic return to the international stage." *Culture, Identity, and Security: An Overview* (New York: Rockefeller Brothers Fund, 1998) p. 4. http://www.rbf.org/resources/resources_list.htm?page_num=2&cat_id=1669

of Civilizations, Amir Pasic of the Rockefeller Brothers Foundation could still write: "Whatever the media may have to say about 'ancient hatreds,' it is a well-established fact that cultural difference by itself is not a leading cause of conflict," adding that "despite Huntington's thesis regarding future fault lines among civilizations, most scholars who study international war do not think culture is an important cause."[24] That "fact" would lose much of its reputation after 9/11, as sales of The Clash proceeded skyward.

Though lacking both the *cachet* of Harvard and the seductive prose peppered with oversimplifications tailor-made for the sound-bite generation, a far better case for the crucial relevance of culture to a proper understanding of world affairs was made by the brilliant professor of international relations Adda B. Bozeman in *Politics and Culture in International Relations: From the Ancient Near East to the Opening of the Modern Age*. In her preface to the second edition of this masterpiece published in 1994, Bozeman presciently observed that "the appellation 'state' tells you nothing nowadays. You may be dealing with a non-state like Somalia; or with a state that serves as cover for anti-state activities or for terrorism; with states that do not subscribe to the principle of territorially bounded space; or that are conditioned by their religious belief systems to be always at war."[25] What her painstakingly researched tome lacks in sexiness, it more than makes up in serious scholarship and profound insight.

The business community would soon discover the importance of cultural knowledge as a necessary tool in a globalizing economy. The complexities of working with people from other environments, with different customs and ways of thinking, had to be understood. Brooks Peterson's *Cultural Intelligence* advertises itself as a guide to do just that: dealing (and wheeling) with foreign publics. The book's title is obviously a take-off from the earlier, more general concept: "emotional intelligence," first proposed by psychologists Peter Salovey and John D. Mayer in 1990, who mapped out in great detail the ways in which emotions may be used more intelligently.[26] After psychologist and New York Times journalist Daniel Goleman popularized the growing scholarship on the subject, the concept instantly took on. Citing Salovey, Goleman identified five main aspects of emotional intelligence: (1) Knowing

---

24 Ibid, p. 26.

25 Adda B. Bozeman, *Politics and Culture in International Relations: From the Ancient Near East to the Opening of the Modern Age*, 2nd edition (New Brunswick, NJ: Transaction Publishers, 1994), p. xli.

26 Peter Salovey and John D. Mayer "Emotional Intelligence," in *Imagination, Cognition and Personality* (1990), pp. 185-211.

one's own emotions (self-awareness); (2) managing emotions; (3) motivating oneself; (4) recognizing emotions in others; and (5) handling relationships.[27] Evidently, people differ in their native abilities to handle one or more of these aspects, but the brain is highly malleable and can adapt considerably. So one is not "programmed" to emotional imbecility; there is room to learn.

Taking it to the next level, Peterson applies this concept to intercultural communication. "Cultural intelligence," he writes, "is the ability to engage in a set of behavior that uses skills (i.e., language or interpersonal skills) and qualities (e.g., tolerance for ambiguity, flexibility) that are tuned appropriately to the culture-based values and attitudes of the people with whom one interacts."[28] Such attitudes are reflected in communications whose inscrutability defies the mere dictionary. When a Chinese and Japanese businessman says "yes" when he means "maybe," "maybe" when he means "no," and "it's difficult" or even "yes" when he means "no," he is not lying. Rather, suggests Peterson, "this represents indirectness, not dishonesty."[29] Such distinctions can make the difference between deal or no deal. In a more ominous context, it could make the difference between war and peace.

Understanding the enemy

For many years, intelligence analysts were taught epistemology (*logos episteme*, the study of how we know) by Richard J. Heuer, Jr., whose articles written during 1978 and 1986 for internal use within the CIA Directorate of Intelligence were republished and edited into a textbook, entitled *Psychology of Intelligence Analysis*, which the CIA subsequently posted online. One of the most common cognitive traps that Heuer exposes in his book is mirror imaging, described as "filling gaps in the analyst's own knowledge by assuming that the other side is likely to act in a certain way because that is how the U.S. would act under similar circumstances."[30] Heuer warns against relying on taking the U.S. perspective on what constitutes another country's national interest, citing by way of example the erroneous assumption in 1977, notwithstanding ample evidence to the contrary, that South Africa would not want a nuclear

---

27 Daniel Goleman, *Emotional Intelligence:Why it Can Matter More than IQ* (New York: Bantam Books, 1995), pp. 46-47.

28 Brooks Peterson, *Cultural Intelligence:A Guide to Working with People from Other Cultures* (Boston, 2004: Intercultural Press, Nicholas Brealey Publishing), p. 89.

29 Ibid, p. 117.

30 Richard J. Heuer, Jr., *Psychology of Intelligence Analysis* (CIA), p. 70.

weapon. Similarly, Adm. David Jeremiah blamed the Intelligence Community's failure to predict India's nuclear weapons testing on this fallacy, calling it the "everybody-thinks-like-us mind-set."[31]

Not that we should distrust all our gut feelings; sometimes, indeed often, the most reliable source of knowledge is subconscious. The runaway bestseller *Blink: The Power of Thinking Without Thinking*, by the talented *New Yorker* journalist Malcolm Gladwell, recasts the ancient truth that we know far more than we think we know. In other words, not only is conscious "knowledge" the mere tip of an impressive iceberg, that tip may be less trustworthy than what lies submerged. At the very least, the latter can influence the former in astonishing ways.

This is both good news and bad. Given the myriad different decisions that we must make with little relevant information, it is reassuring that evolution has equipped the human race with fairly reliable instincts. But trouble arises when people forget their own limitations. It is important to remember, for example, that early experiences inevitably color later perceptions and may significantly skew analysis. Inaccurate racial and other "profiling" judgments are also far more prevalent than we would like to believe, as prejudices and stereotypes are often triggered unconsciously.[32] And in wartime, such epistemological flaws can prove catastrophic.

Fortunately, recognizing the problem is half-way to solving it. Several studies indicate that prejudices may be addressed with considerable success.[33] In other words, epistemology is not destiny. But if all knowledge is conditioned by experience, it follows that culture plays a huge role. And the cultural dimensions of human judgment are critical – again, for good and ill. There is a reason why someone who grew up in a particular milieu is better equipped to understand the many subliminal forms of communication in that milieu than a total stranger. That said, it is possible to overestimate one's own knowledge and miss signs that may be immediately obvious to an observer unhampered by overconfidence.

Socrates was right to ask questions; wise is he who emulates the great philosopher. Answers are invariably trickier than anticipated. Who, for instance, counts as an enemy? A friend? Is it always easy to tell them apart? What turns an enemy into a friend – and, conversely, a friend into an enemy?

31 Jim Wolf, "CIA Inquest Finds US Missed Indian 'Mindset,'" UPI wire service, June 3, 1998.

32 Malcolm Gladwell, *Blink: The Power of Thinking Without Thinking* (New York: Little, Brown & Co., 2005), ch. 3.

33 Several sources are cited by Gladwell on p. 258.

Is there a continuum? Indeed, isn't everyone in some ways his own enemy – sometimes even knowingly, as if paradoxically being driven by a "death wish"? Addressing such queries is critical to an understanding of any battlefield, both literal and ideological, and indispensable for winning not only the war on the ground but the war of ideas. Without both knowing and adequately taking into account the cultural terrain, winning the peace is unthinkable, no matter how impressive and overwhelming the military acumen and hardware.

Perhaps the most celebrated instance of a superior force winning a war yet losing the peace as a result of fatally underestimating the significance of what has come to be known as "the human terrain" was Napoleon's failure, two centuries ago, to influence positively the populations of Spain and Portugal in the aftermath of his preemptive occupation. The many lessons provided by that failure have mostly gone unheeded. Lt. Col. George W. Smith, Jr. warns:

> These lessons demonstrate that an inordinate focus on armies at the expense of a focus on the people has and will continue to make winning the peace more difficult than winning the war. Closing the cultural intelligence gap by striking an IPB balance within campaign planning may reduce surprises for an occupying force that historically have impeded the accomplishment of the campaign's stated political or grand strategic objectives.[34]

This is especially relevant in a situation of hybrid warfare, where "armies" is not the most appropriate way to describe the forces required to defeat so-called insurgents, various Islamist and other extremist guerilla-style irregular forces. Ethnocentrism has been one of the main obstacles to successful strategic engagement. It may well be argued that an absence of cultural acumen makes it harder to prevent unnecessary violent conflict. But even when military force must be used, it should be wielded deftly. No warrior can afford to ignore the principal lesson of the great Sun Tsu, who wisely declared two millennia ago: "If you know the enemy and know yourself, you need not fear the result of a hundred battles."[35] He could have added that if you are lacking in both, a hundred battles will not suffice: you will still lose.

Anthropologist Montgomery McFate reminds us that the United

---

34 Lt. Col. George W. Smith, Jr. "Avoiding a Napoleonic Ulcer: Bridging the Gap of Cultural Intelligence (Or, Have We Focused on the Wrong Transformation?),
http://www.au.af.mil/au/awc/awcgate/usmc/cjcs_essay_smith.pdf. Also see chapter six of this book.

35 Sun Tsu, *The Art of War.*

States is facing an enemy that is typically non-Western, transnational, non-hierarchical in structure, operating outside the nation-state, whose "form of warfare, organizational structure, and motivations are determined by the society and the culture from which they come. Attacks on coalition troops in the Sunni triangle, for example, follow predictable patterns of tribal warfare."[36] It is now a commonplace that the Bush administration misunderstood the tribal nature of Iraqi culture and society, having assumed that the civilian apparatus of the government would remain after the regime had been destroyed by air strikes.

The U.S. also misunderstood the system of information transmission in Iraqi society, worrying more about anti-coalition sentiments in the press, quick to close newspapers and destroy the Iraqi electronic communication infrastructure, all the while preventing soldiers and marines from establishing one-to-one relationships with Iraqis, "which are key to both intelligence collection and winning hearts and minds." For example, American troops were alarmed by all the gesticulating and agitation. Noted one Marine: "We had to train ourselves that this was not threatening. But we had our fingers on the trigger all the time because they were yelling." A lack of familiarity with local cultural symbols also created problems. For example, in the Western European tradition, a white flag means surrender. Many Marines assumed a black flag was the opposite of surrender—"a big sign that said shoot here!" as one officer pointed out. As a result, many Shia who traditionally fly black flags from their homes as a religious symbol were identified as the enemy and shot at unnecessarily."[37]

Half a decade into the conflict, Americans have learned a great deal. Cultural Studies courses are being offered throughout the educational system of the Defense Department. The essay by LTC Fred Renzi, for example, has made a strong case for "ethnographic intelligence," which is the main purpose of the so-called HTS (Human Terrain System) – "in reality a CORDS project for the 21st Century."[38] CORDS is the acronym for Civil Operations and Revolutionary Development Support program, a project administered jointly

---

36 Montgomery McFate, "The Military Utility of Understanding Adversary Culture," *Joint Force Quarterly*, July, 2005 www.dtic.mil/doctrine/jel/jfq_pubs/1038.pdf. Also see chapter seven of this book.

37 Ibid.

38 LTC Fred Renzi, U.S. Army, "Networds: Terra Incognita and the Case for Ethnographic Intelligence," *Military Review*, Sept-Oct 2006., www.army.mil/professionalwriting/volumes/volume4/december_2006/12_06_1.html. Also see chapter nine of this book.

by the South Vietnamese Government and the Military Assistance Command Vietnam (MACV). Implemented under the Johnson administration, the CORDS program specifically matched focused intelligence collection with direct action and integrated synchronized activities aimed at winning the "hearts and minds" of the South Vietnamese. Writes Renzi: "CORDS was premised on a belief that the war would be ultimately won or lost not on the battlefield, but in the struggle for the loyalty of the people."[39]

Former CIA Director William Colby later blamed the final loss in Vietnam on failure to implement the CORDS strategy, asserting that the "major error of the Americans in Vietnam was insisting upon fighting an American style military war against an enemy who, through the early years of the war, was fighting his style of people's war at the level of the population."[40] Specifically, he charged that efforts to transform rural life through economic development would have created the conditions necessary to foster peace and stability. Such development, he maintained, would have countered any appeal the terrorists might have had for the population by creating local opportunities for the people to exercise real freedoms within their own institutions and values – an assessment later confirmed by a number of studies mentioned by Renzi. HTS is designed to apply precisely these kinds of lessons.

But we have much catching up to do – especially as history refuses to stand still. No sooner has the term "asymmetrical warfare" been introduced to describe engagement with a weaker adversary using unconventional stratagems, weapons, and other capabilities against us, the strongest military power in the world, than it all but lost relevance. As John W. Jandora points out, after 9/11 the form of asymmetric warfare that has tormented the Western military establishment has been militant jihadism, which requires a panoply of novel strategies and approaches.

"Jihadism," to be sure, is by no means monolithic, nor should the U.S. rule out the possibility that some Muslim state may help us against what Jandora calls "the jihadists' center of gravity." It is certainly too soon to predict which Muslim state might ultimately play such a role, let alone whether the Defense and State Departments would be able to sit down and talk long enough

---

39 Jacob Kipp, Lester Grau, Karl Prinslow, and Cpt. Don Smith, "The Human Terrain System: A CORDS for the 21st Century," *Military Review*, Sept-Oct 2006, www.army.mil/professionalwriting/volumes/volume4/december_2006/12_06_2_pf.html. Also see chapter ten of this book.

40 William Colby with James McCargar, *Lost Victory: A First-Hand Account of America's Sixteen Year War in Vietnam* (Chicago & New York: Contemporary Books, 1989), 175-192.

to agree on a national strategy. Jandora is convinced, however, that "regardless of which government leads and whether the requisite interagency approach ever becomes reality, the U.S. military must prepare to factor culture into mission planning at tactical, operational, and strategic levels."[41]

One size does *not* fit all

The "what's in-what's out" mind-set seems to have infected both popular and military culture: we are enamored of fads and fashion. A senior PSYOP (psychological operations) official refers to "human terrain" as the latest flavor-of-the-month in U.S. strategic training. This doesn't imply merely that it may not last; it can also mean that lessons from one region are applied without requisite sophistication to another, where they may be peripheral or even counterproductive. Commonly known as "the cookie cutter approach," the outcome can be serious strategic indigestion.

That is precisely what MAJ Christopher Varhola and LTC Laura Varhola warn against. It seems that we are so obsessed with the need to train people for operating in the Middle East that we ignore other parts of the world. For example, US forces bound for East Africa, prior to deployment received training for – you guessed it - Iran and Afghanistan. The lack of regional training and overall expertise contributes to further isolating US forces, preventing them for adequately integrating into foreign societies. To make matters worse, "they sometimes reside in luxury hotels and hire translators or 'expeditors' to procure items in the local economy and to advise them on how to interact with locals. Sustained operations have involved the creation of luxurious "safe houses" in the wealthy expatriate communities of East Africa."[42] While such an arrangement may meet embassy guidelines for force protection, it evidently prevents American troops from learning about the country, to say nothing of generating understandable local resentment.

Nor is that all. As often happens, especially in America, short-term trumps long-term, and tactical trumps strategic. Recognizing that culture matters on the battlefield -both during hot war and the subsequent quest for

---

41 John W. Jandora, "Center of Gravity and. Asymmetric Conflict. Factoring. In Culture," *Joint Force Quarterly*. Oct 2005, www.dtic.mil/doctrine/jel/jfq_pubs/1439.pdf. Also see chapter eight of this book.

42 MAJ Christopher H. Varhola and LTC Col. Laura R. Varhola, "Avoiding the Cookie-Cutter Approach to Culture:. Lessons Learned from Operations in East Africa," *Military Review*, November-December 2006, www.usacac.army.mil/CAC/milreview/English/NovDec06/indexnovdec06.pdf. Also see chapter eleven of this book.

peace - is a good first step, but it doesn't go far enough. Actionable knowledge requires a deep and sophisticated understanding of both tradition and evolution. As Sheila Myioshi Jager writes in her recent thesis "On the Uses of Cultural Knowledge," we must "take into account the vital role of history and historical memory. Culture is not unchanging, nor does it entail a set of enduring values and/or ancient 'patterns' of thought from which we can predict behavior."[43] Useful as it surely is to study patterns of culture as taught by an academic specialist in a war college setting, strategists need more.

Writes Jager:

> Cultural knowledge at this level thus requires a complex understanding of culture as a dynamic entity, an on-going process of negotiation between past and present. Far from reproducing the values and beliefs of a static and unchanging culture, extremist groups like Al Qaeda have appropriated and reinterpreted Islamic texts, belief-systems, and traditions to justify their own radical ideology; in other words, they have used culture instrumentally. Cultural knowledge as applied to the level of strategy must be concerned with the dynamic understanding of culture and how different Islamic radicals emphasize different aspects of their historical past and traditions to legitimize their political actions and behavior in the present.[44]

This requires different methodologies, and a serious long-term commitment to addressing the challenge on many levels. Peace might be won in the short term with less, but the price may be high. Vigilance requires steady attention to detail, and the United States should be able to rise to the occasion. Arguably, it cannot afford not to.

Among such details is the emerging significance of female warriors among Muslim groups traditionally hostile to women's role in any public capacity. In 2004, for example, the extremist anti-Israeli Islamist group Hamas reversed its original opposition to using women suicide bombers. Debra Zedalis explains how using women as suicide attackers can provide some obvious benefits:

• Tactical advantage: stealthier attack, element of surprise, hesitancy to search women, female stereotype (e.g., nonviolent).
• Increased number of combatants.
• Increased publicity (greater publicity = larger number of recruits).

---

43   Sheila Miyoshi Jager, "On the Uses of Cultural Knowledge," Strategic Studies Institute, November 2007, www.strategicstudiesinstitute.army.mil/pdffiles/pub817.pdf. Also see chapter twelve of this book.

44   Ibid.

• Psychological effect.[45]

"It is the ultimate asymmetric weapon," comments Magnus Ranstorp, director of the Center for the Study of Terrorism and Political Violence. "You can assimilate among the people and then attack with an element of surprise that has an incredible and devastating shock value."[46] Yet despite its increasing popularity, the use of women as weapons is still insufficiently understood by Western analysts.

No compilation of readings on cultural intelligence, however, would be complete without taking note of the ingenuity exhibited by military personnel on the ground when faced with the need to operate in situations for which they had been unprepared. A shining example is BG Ralph O. Baker, U.S. Army, who was commissioned by *Military Review* to write about his experiences after his return as commander of the 2d Brigade Combat Team of the 1st Armored Division in a volatile area in Baghdad.

BG Baker candidly admits that at first, the incredibly complex cultural terrain threw him for a loop. Having shared the bias of his fellow soldiers against "information operations," he soon learned the hard truth that winning in such a climate would be impossible without local cooperation and took the lesson to heart:

> We identified the key leaders in our AO who wielded the greatest influence. These included clerics (Sunni and Shiite imams and Christian priests from Eastern Orthodox churches), sheiks and tribal leaders, staff and faculty at the universities (a group that has incredible influence over the young minds of college-age students), local government officials whom we were mentoring, and finally, select Arab media correspondents. We began our leader engagement strategy by contacting members of local governments at neighborhood, district, and city council meetings. We sat side by side with elected local council leaders and helped them develop their democratic council systems. Eventually, we took a backseat and became mere observers. My commanders and I used these occasions to cultivate relationships with the leaders and to deliver our talking points (never missing an opportunity to communicate our two brigade themes).[47]

45 Debra D. Zedalis, "Female Suicide Bombers," Strategic Studies Institute, June 2004, www.strategicstudiesinstitute.army.mil/pdffiles/PUB408.pdf. Also see chapter thirteen of this book.

46 See Don Van Natta, Jr., "Big Bang Theory: The Terror Industry Fields Its Ultimate Weapon," *New York Times*, August 24, 2003, sec. 4, p. 1.

47 BG Ralph O. Baker, U.S. Army, "The Decisive Weapon: A Brigade Combat Team Commander's Perspective on Information Operations," *Military Review*, May-June 2006, www.usacac.army.mil/CAC/milreview/English/MarApr07/Baker.pdf. Also see chapter

The dramatic experiences of BG Baker were repeated elsewhere in both Iraq and Afghanistan – indeed, throughout the world, as Americans serving their nation in difficult circumstances rose to the occasion, learning about the people around them and seeking to help them while simultaneously advancing the strategic goals of the U.S. The results have been remarkable. Among the most memorable public accounts is Robert D. Kaplan's bestselling *Imperial Grunts: The American Military on the Ground*. Here's a scene from a village near the Afghan border with Pakistan, where there had just been fighting between the Mangal and Totakhel Pushtuns:

> You could tell if a village was friendly by the behavior of the children: if they came out and waved you were welcome; if they didn't you weren't. .... Special Forces troops always brought along extra Power Bars for the kids on these trips and took pictures with them – anything they could think of to break down barriers... There was much chai consumption. Intelligence was best gathered not by asking direct questions, but simply by establishing relationships..... The provincial reconstruction teams [PRTs] were a new and trendy concept in the autumn of 2003. The PRTs did the very things media loved: civil and humanitarian affairs – nation-building, that is. The PRTs were inter-agency, combining different military units and governmental departments into a single package. They constituted a recognition that in war zones the military, rather than the civilian charities (nongovernmental organizations [NGOs]), was best positioned to carry out civil affairs. The PRTs, while part of the conventional Army, represented a form of unconventional warfare, for humanitarian aid, besides winning 'heart and minds,' and thus breaking the link between the insurgents and the general population, was a useful cover for informal intelligence gathering. The global media was less comfortable with this aspect of PRTs.[48]

Global media sensibilities aside, if it weren't dangerous enough for humanitarian NGOs to operate in hellish environments, being suspected of collaborating with intelligence agencies could render their assistance virtually impossible. It does no good to minimize the dilemmas of conducting such complicated military operations, trying to destroy the enemy while also insuring the safety of those who seek to bring peace to a traumatized community. For starters, the various U.S. government agencies must penetrate barriers inside their own respective cultures so as to more effectively address the challenges of protecting national security.

---

fourteen of this book.

48   Robert D. Kaplan, *Imperial Grunts: The American Military on the Ground* (New York, Random House, 2005), pp. 216-217.

Culture and strategy

The principal relevant U.S. agencies – Defense, Homeland Security, and State – are hardly a model of coordination. Without strong guidance from the White House, moreover, information synchronization is impossible. Writes COL Gregory Julian:

> While there are signs of progress, the DoD efforts in strategic communication currently lack adequate strategic direction, and interagency coordination. The accomplishment of military objectives is at great risk, as well as the potential for squandering national resources in costly military campaigns without a coherent communications strategy. There must be a top-down strategy from the President to synchronize efforts of diplomacy and public information with activities of various other elements of national power to support foreign policy objectives. The information campaign must be a continuous process in peacetime, during military campaigns, and throughout stability or peacekeeping operations.[49]

Winning the peace is proving to be far harder than fighting hot wars. America's strategic structure is deeply flawed, with little indication that improvement can be expected anytime soon.

Meanwhile, Islamism continues to present a formidable threat to the only superpower left, a de facto empire that is still trying to come to terms with the nature of its adversary while fighting internal demons of its own making. John J. Dziak explains:

> In responding to a resurgent Islamism that is prosecuting a new generation of warfare especially in its deceptive dimensions, U.S. intelligence faces several difficulties, not all of them within its powers to address. With regard to the latter, public opinion, especially in its elite manifestations, is conflicted about linking a major world faith with terrorism and other violence. Coupled with fixations with political correctness and widespread nonjudgmental relativism, strong elements of U.S. opinion seem still to be in a pre-9/11 mindset. There is a palpable reluctance in many quarters to admit that a conflict along civilizational/cultural boundaries is underway. A deep reluctance to acknowledge this is seen in the censored language of public discourse, i.e., the use of generic "terrorism" vice Islamists, radical Islam, etc.[50]

By way of confirmation, the Department of Homeland Security's Office for Civil Rights and Civil Liberties issued guidelines "For Official Use

---

49 COL Gregory Julian, United States Army, "Transforming the Department of Defense Strategic Communication Strategy," US Army War College, March 15, 2006, www.strategic-studiesinstitute.army.mil/pdffiles/ksil386.pdf. Also see chapter sixteen of this book.

50 John J. Dziak, "Islamism and Stratagem," *The Intelligencer*, April 6th, 2007. Also see chapter seventeen of this book.

Only" in January 2008, but widely publicized a few months later. They had been allegedly prompted by Muslim leaders' concern that journalists, pundits, and officials, should use language that is "properly calibrated to diminish the recruitment efforts of extremists who argue that the West is at war with Islam." Debates pro and con this semantic restraint notwithstanding, one surely has to question the wisdom of substituting "progress" for "liberty."[51]

Questions about tactics and strategy abound. What kind of "warfare" are we fighting? How do we win "hearts and minds"? How do we not just win but actually keep the peace? These are the hard challenges not only for the West: at stake is mankind's very survival. Perhaps the so-called fourth generation warfare will soon be superseded by a fifth, which Col. T. X. Hammes predicts "will result from the continued shift of political and social loyalties to causes rather than nations. It will be marked by the increasing power of smaller and smaller entities and the explosion of biotechnology."[52] In the event of such a scenario, cultural intelligence will be more critical than ever. And as ideas will continue to matter, the importance of understanding cultures will grow.

Even strategists who take exception to describing the current predicament of the West as engagement in a new generation of warfare, notably Antulio J. Echevarria II,[53] agree that U.S. intelligence should pay a great deal more attention to cultural considerations. Echevarria certainly commends the use of anthropologists in warfare: "The U.S. Army's new Human Terrain System, which helps enhance cultural awareness, is an important step in the right direction and should be supported," he writes. "By developing an understanding of wars of ideas as a mode of conflict, we can fight the current battle of ideas more effectively while at the same time better prepare ourselves to wage future ones."[54] He might have added that the battle goes on even when it seems to have stopped. In other words, we are engaged in a truly "long" war, not merely a "cold" one with occasional hot spots in the form of terrorist attacks. As weapons of mass destruction, both nuclear and biological or chemical, proliferate among irresponsible and evil regimes and individuals, whatever we choose to call it, it will take all the intelligence we can muster.

51   Bret Stephens, "Homeland Security Newspeak," *Wall Street Journal*, May 27, 2008, http://online.wsj.com/public/resources/documents/gloview.pdf.

52   Col. T. X. Hammes, USMC, Retired, "Fourth Generation Warfare Evolves, Fifth Emerges," *Military Review*, May-June 2007. Also see chapter eighteen of this book.

53   See his "Fourth-Generation War and Other Myths," Strategic Studies Institute, Nov. 2005.

54   "Wars of Ideas and the War of Ideas," by Antulio J. Echevarria II, Strategic Studies Institute, June 2008. http://www.strategicstudiesinstitute.army.mil/pubs/display.cfm?pubID=866. Also see chapter nineteen of this book.

Aside from national threats, which have never disappeared – notwithstanding the breathtakingly rapid (and relatively peaceful, at least in the short run) demise of our admittedly most formidable adversary, the USSR - the new enemy is more insidious and challenging than we ever could have anticipated in our premature victory dance at the end of the Cold War. It was undoubtedly in order to set the stage for U.S. policy regardless of who would be President in 2009 that Defense Secretary Robert Gates released a National Defense Strategy in mid-2008, its ongoing relevance enhanced by his continuing as Secretary of Defense under President Obama. That Strategy describes the situation starkly:

> We face a global struggle. Like communism and fascism before it, extremist ideology has transnational pretensions, and like its secular antecedents, it draws adherents from around the world. The vision it offers is in opposition to globalization and the expansion of freedom it brings. Paradoxically, violent   extremist movements use the very instruments of globalization – the unfettered flow of information and ideas, goods and services, capital, people, and technology – that they claim to reject to further their goals. Although driven by this transnational ideology, our adversaries themselves are, in fact, a collection of regional and local extremist groups. Regional and local grievances help fuel the conflict, and it thrives in ungoverned, under-governed, and mis-governed areas. This conflict is a prolonged irregular campaign, a violent struggle for legitimacy and influence over the population. The use of force plays a role, yet military efforts to capture or kill terrorists are likely to be subordinate to measures to promote local participation in government and economic programs to spur development, as well as efforts to understand and address the grievances that often lie at the heart of insurgencies.[55]

In brief, unless conflict is understood as the permanent state of mankind, which requires eternal vigilance and an ongoing effort to both understand and communicate effectively with the rest of the world, the United States of America will flounder in its efforts to stave off the forces of destruction which are more insidious and lethal than ever. The principal prerequisite for a national security strategy is solid, actionable intelligence – which clearly must be predicated on long-term engagement and a realistic assessment of the multiple dangers confronting the global community. For that, however, we must redefine peace as the challenge of survival, this side of Eden.

---

55 National Security Strategy, June 2008, pp. 7-8, www.defenselink.mil/news/2008%20 National%20Defense%20Strategy.pdf

I
Culture

2

War and the Clash of Ideas

Adda Bozeman

"The War of All Against All" is the title of an analytical review of papers that were published in the Journal of Conflict Resolution from 1957 to 1968.[1] Within the protective covers of these volumes, contributors contend for different causes; yet the clash of their ideas is significantly muffled by basic accord on two great issues. The scholars are at one, the reviewer notes, in regarding international war as the category of central interest, and they are united also in stressing conflict control rather than conflict itself. Moreover, they are found to be nearly unanimous in assuming that violence is something to be avoided if at all possible, and in attaching connotations of illegitimacy to the phrase "organized violence."

Given these shared dispositions, it is not surprising to learn, then, that arms control is a heavily favored research subject and that the literature on this topic is pervaded by several common impulses, among them the following: repugnance for "untraditional" methods of warfare or for weapons "which a given nation has not yet had a chance either to arm itself with or to develop counter-weapons against;"[2] disdain for "sham bargaining" and psychological warfare, the latter generally being viewed as sneaky and immoral; the strong conviction that humane-ness ought to be accepted as an important criterion in the evaluation of weaponry; a deep commitment to the distinction between "just" and "unjust" wars; and considerable preoccupation with guilt and responsibility with regard to the actual resort to violence or war.

The same exhaustive survey also instructs us that *JCR* authors have

---

1   Elizabeth Converse, "The War of All Against All: A Review of The Journal of Conflict Resolution, 1957-1968," *Journal of Conflict Resolution* (hereafter cited as JCR), December 1968, pp. 471-532.

2   *Ibid.*, p. 479.

not paid much attention to the relationship between the cause of national survival, on the one hand, and arms control, on the other, and that inquiries into the antecedents of military aggression have been conspicuously absent. In fact, the preferred time dimension has very definitely been the present, amplified by strong overtones of futurist concern. The historical approach is missing, and statistical treatment is stressed; what is more, the data considered relevant to such statistical processing are drawn almost exclusively from American and European records. And finally, it appears that findings by specialists in military science and strategy have been seldom exploited.

Analogous trends have been found to dominate international relations research. Chadwick Alger, in a research review published in 1970,[3] pointed out that concern with the causes of war had given way to study of the causes of peace and the construction of "alternate futures," and that knowledge of the destructive power of nuclear weapons had sparked a revival of interest in disarmament and arms control. He also noted that the peace research movement has been "international in composition, being comprised mainly of North Americans and Western Europeans," that these participants have had high value-commitments to the nonviolent solution of international conflicts and have endeavored to do work with policy relevance, and that they have stressed "scientific work," including systematic data collection techniques and rigorous methods of analysis. Here, however, as in the field of conflict resolution, the demands of rigorous analysis can obviously be satisfied without methodically utilizing data from non-Western societies.

The processes of theory- and model-building that have been perfected in recent years are certainly impressive, and so are many of the actual mental constructions that have issued from these labors. Yet it is questionable whether objective validity can be claimed for much of this work, if only because it is permeated by paradox. It is necessary, then, to note that most scholarly architects profess to be value-neutral social scientists. This is true even though they

3   Chadwick F. Alger, "Trends in International Relations Research: Scope, Theory, Methods, and Relevance," in Norman D. Palmer, editor, *A Design for International Relations Research: Scope, Theory, Methods, and Relevance* (Philadelphia: The American Academy of Political and Social Science, Monograph No. 10, October 1970), pp. 7-28. Also see Philip P. Everts, "Developments and Trends in Peace and Conflict Research, 1965-1971: A Survey of Institutions," *JCR*, December 1972, pp. 477-510. Everts shows (p. 499) that "Peace research itself," "United Nations problems" and "International organization" scored highest as research topics (67%, 66%, 65%, respectively). Cf. Berenice A. Carroll, "Peace Research: The Cult of Power," ibid., pp. 585-616: "If there is any distinguishing common feature among the highly varied works in the field of peace research, it is an avowed commitment to 'peace.'" (p. 599.) Carroll's remarks on "conceptions of power" are particularly interesting.

admit, directly or indirectly through the medium of their accomplishments, that the major motivation for their sustained efforts clearly originates in the compelling force of their own feelings, impulses and values, notably those that feed their hopes for, and images of, a peaceful world society.

Now there is no reason why social scientists should not have values; nor is there any reason that they should not be concerned with the improvement of the lot of man. But in this case, we discover that personal value preferences have not been checked out objectively before they were judged to be appropriate building blocks for theory. More important, perhaps, few modern theorists in the fields of international relations or conflict resolution have bothered to explore the value content of conflict, war and violence. The configuration of the enemy they profess to fight is thus not clearly rendered – a circumstance that may explain why this "war against war" can be perceived by others as a kind of shadowboxing. Indeed, explicit definitions are missing for both war and peace, perhaps because the ruling supposition is that the one is everywhere known to be the opposite of the other, war being universally disclaimed as a thoroughly bad idea and peace being just as generally accepted as mankind's natural state and birthright.

The clash of ideas over how to control conflict, avoid war and build the structures of peace seems to have proceeded in the calm of an academic environment within which the clash of arms and the clamor of war-affirming rhetoric are not readily heard. Some future nonacademic parliament of man, however, may well entertain the motion that these theorists fiddled while nations burned. Whether an armed conflict today is classified as an insurrection, a civil war, a war of national liberation, a guerrilla war or a war-by-proxy, a UN war to preserve the peace, an international socialist war to serve the cause of revolution, or a traditional interstate war – and the lines of differentiation are becoming increasingly blurred in response precisely to the high incidence of violence and the steady proliferation of types of warfare – the fact remains that the post-1945 world can fairly be viewed as a conglomerate of theaters of war, some self-contained and localized, others contiguous and interdependent.

A simple inventory of bare and incontrovertible facts is revealing: Irish groups seem bent on changing the political order of the island by resort to indiscriminate violence; relations between Israel and the Arab states and peoples have been characterized from 1948 onward by warfare; factional, national and international affairs in the Islamic Middle East have been marked by bloody revolutions, armed interventions, takeovers and ethnic uprisings;

several North African Muslim regimes have consistently warred against the non-Islamic populations to their south; Africa south of the Sahara has been convulsed by interstate and intertribal violence, civil wars, coups d'etat and political assassinations, as well as by military and paramilitary activities on the part of anti-white liberation armies and their Rhodesian and Portuguese opponents; Greeks and Turks cannot resist fighting over Cyprus; the armies of the Soviet Union have crushed numerous national uprisings among the allegedly sovereign states of Eastern Europe; India has chosen force over available peaceful methods in order to establish or retain her dominion over Hyderabad, parts of Kashmir, Goa, Sikkim and such non-selfgoverning territories as those occupied by the Naga hostiles; India and Pakistan have not had any scruples about settling their conflicting claims and interests on the field of battle; there would have been no Bangladesh had there not been ruthless warfare; it was China's armed might that subdued Tibet and successfully asserted control over the Paracel Islands; generations of Koreans have known nothing but the actuality or the threat of civil and international war; the destinies of all peoples in the vast Southeast Asian region have long been molded by war, whether in the form of armed uprisings and revolutions, jigsaw movements of insurgency and counterinsurgency, belligerent confrontations between neighbors, or military interventions by great powers; terrorist organizations of one hue or another operate freely throughout Latin America; the United States, which is the academic center of the search for a warless world, has not only warred against communist forces in Korea and Indochina but is itself the troubled scene of terrorist activities by self-styled liberation armies, urban guerrilla bands and other violence-espousing groups.

War's overwhelming and variegated presence would seem to be at odds with some of the major assumptions relayed explicitly or implicitly by leading theorists of international relations and conflict resolution. Doubt may thus be cast on the proposition that "international war," the category of foremost concern, can be convincingly extricated from the maze of other types of warfare in which modern nations are enmeshed. Likewise, and for the same reasons, it is questionable whether distinctions between combatants and non-combatants, or between humane and inhumane weapons, can be maintained effectively, or whether one can endorse the proposition that clear-cut lines between aggression and defense (and thus between just and unjust wars) are always readily discernible. Specialists in military science have closely studied just such issues; however, as the aforementioned analyses suggest, their findings do not seem to have had a vital impact on present trends in political science, peace research, arms control or conflict resolution.

Other incongruities between theory and reality are suggested by the raw evidence of modern war and violence. A glance at the embattled and conflict-ridden regions of Africa, Asia, Latin America and parts of Europe leaves one with the strong impression that human dispositions toward stress, violence and death are by no means everywhere the same, and that basic orientations toward war and peace are therefore greatly various also. For example, nowhere outside North America and Northern Europe does one encounter the overriding desire to avoid armed conflict and to seek peaceful settlement of disputes that leading peace-minded scholars in our society assume to be generally present.

Furthermore, evidence is totally missing that recourse to armed force evokes feelings of guilt and self-recrimination among the intellectual elites of non-Western societies, or that the high incidence of organized and unorganized violence induces doubts about the appropriateness of ruling moral or political systems. Indeed, the strife-filled records of the past twenty-five years, together with the conflict-laden language so often employed by spokesmen for African, Asian and communist societies, point to the possibility that conflict and violence may well be accepted in most areas outside the Occidental world as normal incidents of life, legitimate tools of government and foreign-policy-making, and morally sanctioned courses of action.

Propositions such as these have not been thoroughly tested in the laboratories of peace research, perhaps because they relate, in the final analysis, to values; and values may resist the kind of "rigorous analysis" that has been aimed at by scholars. At any rate, it is noteworthy that eminent theorists in the fields here under review have refrained altogether from probing the mental and psychocultural roots of war, that they have not been much interested in the historical antecedents of actual conflict situations, and that they have not thought of war as a complex of possibly quite disparate, even irreconcilable, norms, values and ideas. Just why these matters have not surfaced in the mainstream of their investigations is in itself a significant thematic motif in the clash of ideas detonated by modern warfare, and as such it should be scrutinized before going any further.

II

Several learned commentators on conflict and its resolution have drawn attention to the fact that today's scholars are uneasy in the face of all, not merely armed, conflict. They are inclined to view it negatively – as an unfortunate interruption of the normal flow of social life, a failure in communication, an unregulated and hence possibly illegitimate transaction, or an aberration from patterns of rational behavior that should be and can be reduced, transformed or eliminated because it is situational rather than instrumental, pathological rather than sane.[4] The exact norms, patterns and models against which motives and actions are judged normal or abnormal are not usually set out. The argument in almost every case appears to be that they are generally known or, to put it differently, that we are here in the presence of some universal givens that need only be implied.

Moreover, and in striking contrast to scholars from an earlier time (notably, Georg Simmel and Robert McIver), conflict today is generally not associated with sentiments, values or psychic states of being. The stress is rather on concrete struggles or overt episodes in which individuals or groups contend for tangible rewards. Thus conceived in terms of antagonistic poles representing two or more mutually incompatible positions, conflict is suspect at the very start, for it is presumed to spring from some kind of discord that could have been avoided.

This neglect of psychological and intellectual factors in situations of social stress seems to have attached itself almost automatically to scholarly thought about those international conflicts that fall short of military war, known in history as cold wars or wars of nerves. Here again, the premise is widely accepted today that clashes of ideas are somehow either irrational departures from the ground rules of normal behavior or ruses to cover up peace-defying policies. In either case it seems to be supposed, particularly in so-called revisionist academic circles, that the Cold War between this country and the communist states was somehow officially initiated in much the same way that hot wars have been declared, and that it could therefore be called off by political authorities in an equally expeditious manner. In other words, conflict is presented as a willed event, rather than a process or relationship,

---

4   See Converse. Also Norman A. Bailey, "Toward a Praxeological Theory of Conflict," *ORBIS,* Winter 1968, pp. 1081-1112. Further see Clinton F. Fink, "Some Conceptual Difficulties in the Theory of Social Conflict," *JCR,* December 1968, pp. 412-460. Fink defines social conflict as "any social situation or process in which two or more social entities are linked by at least one form of antagonistic psychological relation or at least one form of antagonistic interaction." (p. 456.) For another flexible approach, see Robert A. LeVine, "Anthropology and the Study of Conflict: Introduction," *JCR,* March 1961, pp. 3-15. LeVine distinguishes among intrafamily, intracommunity, intercommunity and intercultural conflict, and points out that there are pervasive conflict types that spread to several levels.

perhaps in deference to the controlling conviction that "war" and "peace" are always absolutely polarized, mutually exclusive, strictly factual conditions, and that total peace must naturally take over when the fighting stops.

It is difficult to find precedents for this modern, chiefly American orientation toward chronic international discord. The history of Europe, which is very much a history of ideas, and therefore also one of clashing ideas, is replete with such wars of nerves. None has been more protracted or more richly documented than the uneasy coexistence of Christian and Muslim in the lands of the Mediterranean, for which contemporary Spaniards coined the term *guerra fría*. This early model of the clash of ideas in international relations has obviously not been examined by today's schools of peace and conflict studies. Nor have they taken note of the unremitting, politically and intellectually poignant collision of beliefs in the minds of statesmen, scholars and ordinary citizens that was set off by the French Revolution and continued unabated long after the smoke had cleared. There is thus considerable justification for describing this state of affairs as "the case of the missing historian."[5]

Explanations for the absence of this dimension of inquiry may range from a lack of interest in history and doubt about its relevance for future-directed peace research to the premonition that rigorous historical research would not support some of the theorists' most favored visions. And, no doubt, they also include the related inclination to treat each and every conflict as a clearly discernible, and hence definable, factual circumstance that can be undone as quickly and purposively as it has been conjured up.

At any rate, few present-day specialists in conflict resolution seem prepared to associate conflict with mobility and flux, or to think of it as a process not always easily defined or arrested by decisive action. Not many among them, then, would agree with Jessie Bernard, who in 1949 argued that conflict may exist in latent form for years before there is a formulation of issues, a showdown or a crisis. Bernard believed that it is therefore a mistake to limit our thinking about conflict to its overt phase; we should instead accustom ourselves to think of conflict as going on day in, day out in varying degrees of intensity, whether the issues are clearly formulated or not.[6] Yet it is this explication of social conflict, rather than the ultramodern one, that can

---

5 Converse, pp. 476-477.

6 Jessie Bernard, *American Community Behavior: An Analysis of Problems Confronting American Communities Today* (New York: Holt, Rinehart & Winston, 1949), p. 106. For an early commentary critical of conflict theory, see Raymond Mack and Richard C. Snyder, "The Analysis of Social Conflict: Toward an Overview and Synthesis," *JCR*, June 1957, pp. 212-248.

be translated convincingly into the language of international relations to cover that indeterminate continuum of "no war/no peace" commonly known as cold war.

Furthermore, as later sections of this paper suggest, the Bernard concept comes close to explaining the types of discord and disorder most commonly found in the local and international affairs of non-Western societies. Finally, and most important from the humanist's perspective, it captures certain constant motifs in Occidental biography and history, among them the proposition eloquently stated by Ortega y Gasset in his meditations on *Don Quixote*, namely, that life is uneasiness.

The discomfort experienced by many social scientists in the presence of cross-national ideological strife is paralleled by deep apprehension when their thought turns to what, in the language of the trade, is known as international war. Analysts of research trends in disarmament, arms control and peace studies see this reaction as a function of their preoccupation with the awesome specter of nuclear war.[7] This preoccupation is understandable; but the fact remains, first of all, that millions of lives have been extinguished since 1945 not by nuclear weapons but by conventional arms employed in all manner of warfare, terrorism and outright massacre and, second, that political theorists are not nearly as troubled about these actualities as they are about possible future horrors.

To justify their concern, analysts often point to the use of the atomic bomb against Japan in World War II, and their argument is usually heavily encumbered by an insistence on America's "guilt"[8] – an indictment, incidentally, that is seldom softened by the reminder that conventional bombs had in fact visited even greater devastation on some European cities during the same war. Some of this literature thus leaves one with the uncomfortable impression that the fear of that which may be, and feelings of guilt over that which *was*, have come close to paralyzing analysis of that which *is*.

Further reflections on the tangle of sentiments and cerebrations in which so much of our supposedly value-neutral work on war is imbedded confirm this impression. Thus we see that today's intense academic concern about the morality of military operations was activated by the war in Indochina, and not any other past or present war, and that ever since it has expressed itself almost exclusively in revulsion against the war-related policies

---

7   Converse, p. 479; Alger, p. 13.

8   See above, notes 1-4.

of the United States and some of her allies. Nor has this massive volume of accumulated professorial indignation been strained and sifted in an objective, methodical manner in order to salvage those elements germane to theory. In fact, there are indications that the opposite tendency is being favored, in the sense that sentiment is being allowed to drift. For now that the international war in Indochina is officially terminated, and now that it is possible, in virtue of spellbinding legal or moral fictions, to view military activities in Asia as "unofficial" or "illegitimate," scholarly offensives are directed against noncommunist Asian governments, which continue to be embattled.

A group of renowned American experts on East Asian history, government and culture has thus felt justified, "in the name of humanity and human rights," to protest "the injustice and the inhumanity" of certain judicial and administrative measures that South Korea has taken against some of her citizens.[9] Since similarly severe protests have not been lodged against the dictatorships of North Korea, North Vietnam or the People's Republic of China, one can only conclude that some private bias is at work here. In this case, as in others,[10] the "missing historian" is an important factor particularly puzzling here since the charges are formulated with the consent of East Asian historians who must be presumed to know that human rights and civil liberties are not part and parcel of traditional administration in Korea, China or the states of Southeast Asia.

The overwhelming presence of private sentiments and values that one detects in war-related literature today does not favor the refinement of ideas into reliable, universally applicable theories about the place of war in human existence. Yet it is definitely theory that students of international relations, war and conflict want most fervently. Indeed, the search for this type of intellectual certainty has been so ardent and compulsive in recent decades that the nontheorist is left with the intriguing image of war-weary troops of academics beating a hasty retreat – away from the unnerving uncertainty of life on the fields of battle and back to the secure shelters of ideationally perfect castles in the mind. But here the refugees are also faced with most demanding problems. After all, social science theory is best attained today if the number of variables is reduced as starkly as possible and if only readily quantified data are considered. Primary attention is therefore usually directed to specific yet sufficiently simple events that can be counted, compared and categorized with

---

9 "U.S. Urged to Cut South Korean Aid: Group Calls for a Protest on Seoul's 'Inhumanity,'" *New York Times*, July 15, 1974.

10 See the scholarly surveys of the literature on war and conflict cited previously.

relative ease.

Is modern war susceptible to this kind of academic processing? If it is true, as the UNESCO Charter states, that "wars begin in the minds of men" and that "ignorance of each other's ways and lives has been a common cause ... of that suspicion and mistrust between peoples of the world through which their differences have all too often broken into war," should it not follow that one must probe the minds of men in search of all the images, beliefs, sensations, values, concepts and modes of reasoning that relate to war?

The data thus collected would of course be infinitely various as well as precarious – the kind not easily stored in data banks as these are now constructed. For just how does one quantify pride, prestige or prejudice, moral outrage, insistence on survival, vanity and vengeance? What does one do with killing in obedience to spirits of the earth or living ancestors? Where in the theoretician's charts and models is there a place for hatred of the enemy or love of country? Are tools available for a rigorous analysis of self-discipline, cowardice, disaffection or daring? And what are the criteria for an objective, transnational comparison of human inclinations or capacities to inflict violence and sustain war-induced uncertainty, suffering and death? If we have no answers to questions such as these, should we then assume that the meanings of war carried in the minds of the Sudanese and the Bengalis, the Israelis and the Kurds, the Arabs and the Poles, the Hutu and the Greeks are one and the same? Or would it be more prudent not to wonder what men think of war and why they fight?

The latter course seems to be the favored response today in that intense quest for generally valid norms and standards to which priority is being attached. Ultimately, the challenge implicit in the task of theory-building calls for the reduction rather than the addition of variables; and on balance one can say that this challenge has been met. It may well be that the decision to overlook sentiments, beliefs and values – in short, the intangibles that resist quantification – explains why international conflict, including war, is now being treated by so many theorists as a special case of social conflict whose paradigm is economic conflict, the category most amenable to data-processing techniques. This choice of emphasis, again, can be traced to the simple but ruling supposition that the norm for the organization of all societies is the modern industrial society of the West, and that the typical human being is therefore rightly envisaged as a man functioning rationally in such an economic environment.

In this kind of theoretical scheme, Hans Morgenthau explains, nations confront each other not as living historic entities, with all their complexities, but as rational abstractions after the model of "economic man" – playing games of military and diplomatic chess according to a rational calculus that exists nowhere but in the theoretician's mind.[11] Nor is it surprising, in light of such pervasive assumptions, that past and present data from non-Western societies have not been analyzed on their intrinsic merits and that a recent volume containing no substantive references to non-American or non-European manifestations of human conflict could yet be entitled *The Nature of Human Conflict*.[12]

This strong trend to constrict the frames of inquiry for the study of modern war has been reinforced in recent years by the steady impact of other firmly held beliefs: trust in the territorial, democratic nation-state as the prototype for political association everywhere on earth; trust, therefore, in the existence of an organizationally unified world society of essentially equal and analogous political units; and trust in the compelling logic and validity of laws of interstate behavior that assign authoritative meanings to all transactions regarding peace, war, neutrality and conflict resolution.

Now, these propositions have had a rather brief and geographically restricted history. They matured in Europe from about 1648 onward in the vortex, it is interesting to note, of almost continuous war. But there, under the auspices of what later became known as the "modern European states system," they did not carry the fixed, exclusive connotations assigned to them today. International history instructs us, too, that our modern, systematized approaches to war, peace, diplomacy and conflict resolution have no precedents in classical, medieval or Renaissance Europe, or in the traditional realms of Africa and Asia. Finally, the actualities of present world affairs strongly suggest that the supposedly pivotal concepts in international relations – that is, the nation-state, the unified world society and international law are either in need of radical revision or beyond repair, casualties as it were in the endless war of ideas on which life appears to feed.

The implications of this obsolescence for any reasoned view of war can be seen most clearly by concentrating one's analysis on the state. It is doubtful indeed whether we are justified in thinking that the territorially delimited, independent nation-state is still universally accepted as the core

---

11   Hans J. Morgenthau, "International Relations: Quantitative and Qualitative Approaches," in Palmer, p. 70.

12   Elton B. McNeil, editor, *The Nature of Human Conflict* (Englewood Cliffs, N.).: Prentice-Hall, 1965).

norm of political organization and, therefore, as the measure by which one distinguishes different types of violence. Surely, nothing comparable to this particular associational form had existed in pre-seventeenth century Christian Europe (which supplied the framework for the coexistence of numerous, quite disparately fashioned, politically active units), in precolonial Africa south of the Sahara, in the Arab/Islamic realm, or in the different civilizations of South, East and Central Asia. Lines of demarcation between local and international milieus of conflict, and between internal and international warfare, were not clearly drawn in traditional societies before the installation, from the nineteenth century onward, of nation-states and legal systems modeled on Western prototypes.

These unifying grafts have atrophied in recent decades under the impact of the following developments: the waning of Western influence and power; the reactivation, in the Orient and Africa, of older, locally respected focuses of authority and communal solidarity; and the successful diffusion of communist doctrines of statecraft, in the context of which the "bourgeois" state is appreciated as a tactical device rather than a value or norm. In short, the concept of the state as a sovereign community, unified politically, morally and territorially, is being subjected to processes of erosion in all parts of the world – not excluding Western Europe and North America. Its substance is being worn away by fragmentation and separatism along narrow ethnic or linguistic lines; by civil disobedience and a faltering faith in law; and by internal war, covert foreign interventions or military aggression from without.

Singly or in combination, these trends account for the dismantlement, division or satellitization of numerous formerly unified and independent polities, on the one hand, and for the creation of new, fully operational political units, on the other hand, which are antithetical to the state in terms of both intention and activity. This is true, for example, of the national and international liberation front, the "provisional" government that functions year in, year out, or the "independent national authority" – the latter a Middle Eastern guerrilla term denoting the embryo of a future Palestinian state. Each of these organizational types is mobile and fluid in the sense that it has no fixed territorial boundaries and no determinate human substance. Furthermore, each exists in virtue of its commitment to violence and war.

The term "international war," then, no longer refers exclusively to violent conflicts between states. Rather, as suggested earlier, it now stands also for a broad spectrum of armed belligerence within the state, ranging from sporadic urban guerrilla activities to civil wars, wars of liberation and secession, insurrections and other revolutionary uprisings, many of which are initiated and maintained in behalf of causes espoused by foreign principals. Moreover, this interpenetration of the domestic and foreign environments effaces altogether the conventionally accepted lines between legitimate and illegitimate force, and puts in question the theoretically established distinctions between war and peace. These interlocking conditions support the conclusion that the state, having forfeited important controlling functions customarily ascribed to it in world affairs, can no longer be regarded as a reliable medium for realistic differentiation among types of war and between the conditions of war and peace.

Next, the erosion of the state as the fundamental, shared norm of political organization, together with general acquiescence in the coexistence of states and anti-state bodies as equal actors in foreign policy arenas, has gradually but ineluctably led also to the devaluation of the two state-based superstructures that provide the context for official foreign relations: (1) the world society of sovereign, equal states and (2) the law of nations, which stipulates the rights and obligations of these states.

Theoreticians in the field of war and peace studies have made scant allowance for these revolutionary developments. Some proceed as if the situation had not changed at all in past decades, while others, heartened by a belief in progress, retreat into the security of self-made legal and political systems that will be actual, they think, in the future. For as the late Martin Wight notes in his essay "Why Is There No International Theory?" the conviction usually precedes the evidence in progressivist international theories. "And when the conviction is analysed or disintegrates," he continues, "one is apt to find at the centre of it what might be called the argument from desperation." In modern times, Wight suggests, this may well be the fear of nuclear war.[13] The argument that the hydrogen bomb has made war impossible thus usually contains two propositions: first, war waged with the new weapons will destroy civilization; second, it is therefore too horrible to happen.

Thus confined, discussion cannot move on either to international actualities or to history, where corrective evidence is readily available. For example, the indisputable fact that the flood tide of modern nonnuclear war has washed away the categories reserved for it by international law and the UN Charter is seldom, if ever, recognized by theorists. Likewise, belief in

---

13 Martin Wight, "Why Is There No International Theory?" in Herbert Butterfield and Martin Wight, editors, *Diplomatic Investigations: Essays in the Theory of International Politics* (London: Allen & Unwin, 1966), pp. 17-34.

the polarity of war and peace is still widespread, even in policymaking circles, and many thoughtful men believe that every war must end — a proposition negated by the reality of continuous armed struggles. Furthermore, it is astonishing that international theorists, notably those committed to the cause of international law, see no purpose in consulting the records of diplomatic and intellectual history.

The international environment to which American and European theorists address themselves today is certainly more vast and diversified than that of either the seventeenth or nineteenth century, when Grotius and Clausewitz, respectively, reflected on the world. And yet, a comparative study of theories then and now leaves the definite impression that war was both being perceived more keenly and explained more accurately by earlier observers, and that the major findings registered in the seventeenth and nineteenth centuries are in harmony with today's reality, whereas those set out most recently are not.

Grotius, writing in a time when the outlines of the modern European states system were becoming apparent, concluded from his reflections on classical, Jewish and Christian thought and action that war per se is not condemned either by the voluntary law of nations or the law of nature, that states may well reduce each other to subjection, that the boundaries of states, kingdoms, nations or cities can often be settled by the laws of war, that wars must employ force and terror as their most proper agents, and that the arguments in favor of war are as numerous as those for the rule of law. "For where the power of law ceases," he writes, "there war begins."[14]

Enduring international peace, by contrast, is presented by this pioneering theorist of international law as a remote condition. The prophesy of Isaiah that the time shall come when "nations shall beat their swords into plowshares, and turn their spears into pruning hooks," when "nation shall not lift up sword against nation" nor "learn war any more," is in Grotius' opinion (as in that of the Jewish prophet) irrelevant insofar as the justice of war is concerned. In the Grotian perspective, the passage merely describes the state of the world that will result if all nations would submit to the law of Christ. Pending consummation of this utopian dream, peace is perforce limited in time and space.

In fact, a significant passage in *De jure belli ac pacis* suggests that it may not always be easy to distinguish between war and peace. War, Grotius notes,

---

14 Hugo Grotius, *The Rights of War and Peace, Including the Law of Nature and of Nations* (Washington: Walter Dunne, Publisher, 1901); translated by A. C. Campbell; Book I, Chap. 2, pp. 4, 8; Book II, Chap. 1, p. 2; Book III: Chap. 1, p. 6; Chap. 8, p. 1; Chap. 21, p. 1; Chap. 25, p. 1.

is a term for a situation that can exist even when warlike operations are not being carried on. Belligerent powers may agree on a cease-fire or truce in the course of war, and no period need be fixed for the continuance of such an arrangement, described by one of his classical authorities as "a transitory peace, in travail with war." "And I shall add," Grotius writes, "that [truces] are made too for years, twenty, thirty, forty, even a hundred years!"[15] In other words, a state of belligerency may well be semi-permanent or protracted.

Theorists after Grotius held rather steadfastly to his major axioms. Clausewitz, whose work On War laid the basis (in the Occidental world of thought) for the systematic study of war as a field of human knowledge, thus restated Grotius when he defined war as the conduct of political intercourse by other means, a form of human enterprise belonging to social existence, and a conflict of great interests that is settled by bloodshed. But he also inveighed against the folly of viewing war as an act of unrestrained violence, a mere passion for daring and winning, or "an independent thing in itself." To Clausewitz, it was quite clear that war is a serious means to a serious end, only a part of political intercourse, and therefore always subject to the political design. And this design, whether understood as referring to a particular foreign policy or to the realm of politics in general, is here decidedly not being viewed as "war by other means" – a theoretical construct in communist conflict doctrine that was to be elaborated several decades later by Lenin, when he stood Clausewitz "on his head."

All histories of diplomacy and the law of nations point to the conclusion that modern Occidental war- and conflict-related thought favors the rule of law and peace. However, they also fully bear out Clausewitz' conclusion: "Peace seldom reigns over all Europe, and never in all quarters of the world."[16]

III

The image of the world that is being rendered today by those social and political scientists with a strong interest in war, peace and conflict resolution is one of a global order of states that are structurally alike in essence or destined to become so under the impact of irresistible leveling forces. In the logic of this tight and finite scheme, all international relations – including

15   Hugo Grotius, The Law of War and Peace: De jure belli ac pacis (New York: Classics Club, 1949); translated by Louise Ropes Loomis; Book III, Chap. 21, p. 1.

16   Karl von Clausewitz, On War (New York: Modern Library, 1943); translated by O. J. Matthijs Jolles; Book I, Chap. 8, p. 57.

belligerent confrontations – are seen as manifestations of national interests that converge on three main unifying themes: the survival of the state, the maintenance of the international system and the avoidance of war. Most of the leading educational texts, syllabuses, and gaming or simulation exercises in the field are therefore elaborations of truths and abstractions that the theoreticians have worked out as if with one mind – and that they therefore seldom question. Thus, since there is no essential difference between State A and State B, there can be none between A's war and B's war.

This explains why conflicts and wars can be added up rather simply to yield some grand total that in turn will point to another universally valid, generally accepted proposition – a process of fact-finding illustrated in the following passage by Robert McNamara:

> In the eight years through late 1966 alone there were no less than 164 internationally significant outbreaks of violence, each of them specifically designed as a serious challenge to the authority or the very existence of the government in question. Eighty-two different governments were directly involved, and what is striking is that only 15 of the 164 significant resorts to violence were military conflicts between two states, and not a single one of the 164 conflicts was a formally declared war. Indeed, there has not been a formal declaration of war anywhere in the world since World War II.
>
> The planet is becoming a more dangerous place to live on not merely because of a potential nuclear holocaust but also because of the large number of *de facto* conflicts and because the trend of such conflicts is growing rather than diminishing. At the beginning of 1958 there were 23 prolonged insurgencies going on around the world. As of February, 1966, there were 40. Further, the total number of outbreaks of violence has increased each year: in 1958 there were 34; in 1965, there were 58.[17]

The exclusive reason for this increase in international violence, we are told, is the obvious fact that so many new states are still economically underdeveloped, a premise evidently no longer subject to verification, as earlier references in this paper have suggested. Again, no allowance is made for the possibility that war-related phenomena are also, perhaps even predominantly, aspects of locally prevalent values, images, traditions and mental constructions. Indeed, explorations of the ways of thought that make or do not make for war, or of the meanings assigned to war and violence in culturally different parts of the world, would quite logically be out of place in the conceptually closed circuit of modern war and peace studies; for how can cultural diversity

---

17   Robert S. McNamara, *The Essence of Security: Reflections in Office* (New York: Harper & Row, 1968), p. 145.

be perceived if "culture" (or "civilization") is not accepted as a relevant variable or factor?

The student embarking on war and peace studies today will look in vain for rigorous analyses of Occidental, Oriental or African philosophies, ideologies, myths and religions. Each volume he consults is likely to contain scores of cross-references to the works of other Western theorists of our era, and scarcely any (in most cases none) to source materials that would tell him how the Chinese or the Indians or the Persians have related to war in the millennia preceding the present moment. Missing, then, are referrals to the writing, for example, of Han Fei Tzu and Mao Tse-tung; to the Mahabharata, which our contemporaries in India continue to read with veneration; to the Koran, which is replete with commentaries on warfare that are eternally relevant for Muslims; or, in the case of Africa south of the Sahara, to the memoirs of modern literate Africans, oral history and the field work of anthropologists.

Anyone interested in uncovering the roots of war-related policies and practices will thus search in vain among today's works on political science or international relations, for access to primary sources is not being stressed anymore. The student in search of authenticity must look elsewhere in the academic universe – notably, it is here suggested, to the humanities, where the uniqueness of men, events and ideas is still recognized and where clashing ideas on war can still be disentangled. Furthermore, in the pursuit of this kind of learning, he may come to accept the world as a "manifold of civilizations" even as he continues to perceive it as a "manifold of states."[18]

"Culture," or "civilization" if one prefers, has been variously defined. Here it will be considered to be all that is fundamental and enduring about the ways of a group; that is to say, it comprises those norms, values, institutions and modes of thinking in a given society that survive change and remain meaningful to successive generations. This point is well illustrated by Paul Verhaegen's discussion of the relation between the "basic psychology" of an African people, on the one hand, and the effects of "cultural transition," on the other. Those characteristics are basic to a culture, he writes, that are dominant in the bush and that remain obvious in even the most Westernized Africans.[19]

---

18 See Harold D. Lasswell, *Psychopathology and Politics*, a new edition with afterthoughts by the author (New York: Viking Press, 1960), p. 240 ff., for a discussion of the state as a "manifold of events."

19 Paul Verhaegen, "Study of the African Personality in the Belgian Congo," in F. R. Wickert, editor, *Readings in African Psychology from French Language Sources* (East Lansing, Mich.: Michigan State University Press, 1967), pp. 242-248.

Similar formulations can be devised for the Islamic realm, notably its Middle Eastern nucleus, India, Southeast Asia, China, Japan and possibly Mongolian Central Asia, including Tibet. Other areas in which distinct norms and values have developed in counterpoint to those brought forth in the West include the communist orbit of the Soviet Union and the Latin American region.

Today, several factors combine in support of civilization as the proper focal point of war research. As preceding comments on the variegated forms of war and violence throughout the modern world have suggested, the Occidental model of the state has ceased to be a reliable indicator or measure of such phenomena as international war and internal war. Indeed, a survey of actually functioning power centers makes it doubtful whether one can still legitimately view the nation-state as the politically controlling, and hence unifying, organizational norm in international relations. Observations such as these, together with reflections on the conspicuous failure of recent American war-related policies, imply rather that we have entered an era in which the interacting, independent units are so disparate that references to an "international order" are invalid. These symptoms of the erosion of the state seem to make it mandatory that we find other or additional ways to determine the configuration of an alien society.

Civilization recommends itself in this respect because it is more comprehensive as an ordering concept than the state: it can cover a host of political formations – armed bands, liberation fronts or empires; anarchies or despotisms; transterritorial commonwealths of commodity producers, financiers or religionists; as well as multinational political parties. Next, also in contrast to the state, a civilization is more enduring in time, even as it is usually less precisely defined in space. And finally, civilization is today a more neutral reference than the state because, contrary to the latter, it is not associated with typically Occidental norms and values. In short, there continues to be great truth in Alfred North Whitehead's remark that a political system is transient and vulnerable by comparison with the principles and forces of the society and culture that have produced it. These principles and forces require explicit recognition before the elements of the political system – in our case, war – can be understood.[20]

It is much harder for Americans than for other peoples to accept such

---

20 For an extended discussion of the nature of civilization and intercultural relations, and of the historical impact of war on the identity of particularly significant cultures, see my "Civilizations Under Stress: Reflections on Cultural Borrowing and Survival,"Virginia Quarterly Review, Winter 1975, pp. 1-18.

a worldview because the United States, almost by definition, stands for the denial of cultural differences and the neglect or irrelevancy of the past. America long ago departed from the European tradition — inaugurated by Herodotus when he explained the Persian Wars as a confrontation between the rival civilizations of Europe and Asia — and today is reluctant to differentiate between wars fought within a culturally unified sphere and those between societies of disparate cultures or idea systems. In fact, after allowance is made for occasional romantic infatuations with insurgencies and wars of liberation in Africa and Asia, it appears that American suspicion of the role of ideas in international relations and foreign-policymaking is so widespread that few wars in either category are accepted as reflecting a clash of ideas.

The voluminous literature on war in the traditional world provides some contrasting perspectives on this age-old human contrivance and, at the same time, yields explanations for the incidence and tolerance of war in each non-Western region. The following brief summaries of culturally and historically basic ideas about war are confined to sub-Saharan Africa, the Middle East, India, Southeast Asia and China.

Sub-Saharan Africa

Since traditional Africa has not produced an organizational form comparable to the Occidental state, "foreign relations" have consisted in interaction among a number of differently organized but self-sufficient units: tribes, clans, villages and other subgroups or divisions. To the extent that so-called empires, hieratic chiefdoms and kingdoms were merely conglomerates of these communities, they were also the scenes of "foreign relations" in which each socially cohesive group was apt to pit itself against the other, even though the "other" would appear to have been part of the "self" from the non-African point of view. This state of affairs, along with the absence of writing and other reliable communications, explains why the radius of intercommunity relations has always been very limited. Furthermore, no widely shared, regionally valid Pan-African institutions for conducting inter-community relations — along the lines of the modern European states system — could develop here, for each small community projected its own social order onto the stage of what we call foreign relations. Black Africa, however, is unified by its culture and a mode of thinking not found elsewhere in the world, and it should therefore not be surprising that we can identify certain uniquely African dispositions with regard to war and peace.

Ethnographers have found that warfare was endemic in all regions of sub-Saharan Africa and that it did not elicit moral qualms. In fact, resort to warfare was logical and necessary in terms of certain deeply held beliefs. War, and organization for war, thus assured the continuous identity of the group as it had coalesced around its own ancestors, origin myths, customs and rites. Moreover, warfare contributed to continual displacements and migrations, resulting in a lack of interest in strictly territorial jurisdiction and thus inhibiting the evolution of a reliable political structure on the order of the European state. Furthermore, war and martial activities embodied the meaning of manhood in tribal life and symbolized the workings of the universe, which was envisioned throughout the continent as the abode of constantly contending, essentially malevolent forces.

Two additional factors need be considered if the role assigned to war and violence in this culture is to be appreciated on its own terms. First, death was not personalized as it is in thought systems that regard the individual as an entity transcending the bounds of the community to which he belongs. Second, death was not objectified as it is in the Western system of causality: in the common African understanding, death was always occasioned by superior, surreal causes, not by a physical weapon; the paramount frame of reference in life was power, particularly magical power, which was associated with ancestral spirits, witchcraft or other supernatural forces.

All traditional structures of African political organization, whether associated with empires, kingdoms, chiefdoms, "anarchies," villages, secret societies or *sub rosa* governments based on fetishism, have been grounded firmly in the view that death is an aspect of society rather than biography, and that conflict, properly staged and manipulated, helps maintain the mythic charter by which a community is ruled. These motifs as well as their organic interaction have found different local expressions, but in certain areas of government – notably the succession to authority and the allocation of power – the separate records converge on a common pattern of institutionalized hostilities, intrigues and internal wars. For example, since it was rare in Africa to find rules that clearly indicated a single heir, succession usually raised rival claimants, resulting in wars for the kingship after an incumbent's death. Whether in the tribal societies of southern Africa, the conquest states of the Interlacustrine Bantus, the kingdoms of the savanna, or among the Mossi and Yoruba in West Africa – just a few of the recognized political systems – ruling circles were rent by quarrels, jealousies and intrigues that were expected to erupt in dynastic, fratricidal or civil wars, and to lead to prolonged periods of anarchy, during which the contest for power would be temporarily resolved.

Not only was this violence often preceded by institution alized regicide, but internal peace did not necessarily follow once the issue of succession had been decided. Since revolts by subordinate princes and chiefs were always expected in East Africa's kingdoms, for example, potential rebels or aspirants to power were routinely murdered or banished. Violent internecine conflict was customary also in Nuba country; among the Nuer, the Kamba, the Masai, the Nandi and other East African peoples; as well as among such territorial groups as the Zulu, the Swazi and the Barotse in southern Africa. Likewise, war was waged regularly by the central governments of most of the imperial domains of West Africa, in order to quell unruly behavior on the part of subordinate regimes.

No agreement exists among specialists in African social organization on just what constitutes rebellion, in which circumstances one can speak of civil war, which episode is properly described as a mere raid or which qualifies as full-fledged aggression. There is general agreement, however, on the proposition that peace was not regarded necessary for the maintenance of the inner order in traditional Africa, that conflict was allowed to express itself in violence, and that warfare among component units of a community was accepted as an organic part of the inner law — provided, of course, that it was employed for purposes considered permissible in a given society. But whether the allowable end was cattle, slaves, women, vengeance or punishment, grazing or water rights, aggrandizement, or the allocation or reshuffling of power, the fact remains that violence has been endemic almost everywhere. Sanctioned by value and belief systems, violence provided, in one form or another, the structural principles for the education of men and the administration of society. Indeed, one might justifiably conclude that internal war was more likely to sustain than to disrupt existing organizational schemes.

Relations between socially or tribally united communities reflect the same fundamental dispositions. Military power, even when wielded by formidable armies, was thus always closely associated with magical power; although concrete rewards such as the capture of cattle or slaves were as prized in the extended martial contest as in the limited engagement, it was the sensation of success left by the investment of superior power that mattered most. And success, again, savored of the enjoyment of a situation in which the enemy of the day was slain or routed and his habitat reduced to ruin. That is to say, victory here was not controlled by expectations of permanent aggrandizement, redemption of lost territories, the extension of a way of life or — with a few exceptions – the installation of a moral system. For those who fought, the end of war was war itself.

All this was in strict accordance with the logic of non-literate, essentially behavioral thought, present-centered time concepts, and the spatial characteristics of African societies. Shrewd calculations of advantage are certainly not missing from the historical records, and particular campaigns, such as the nineteenth-century Ashanti wars, which culminated in the siege of Kumasi, are known to have been planned most methodically. But this sort of comprehensive, long-range planning was not the rule, if only because the future was not seen as separate from the present or the past, and because political identity did not depend on territorial boundaries. Thus strategic thinking, if the term is applicable at all, did not aim at the consolidation of victory by rehabilitating devastated areas, integrating conquered peoples or establishing definite frontiers.

The same ways of thought naturally obtained in defeat since the vanquished were at one with the victors in their basic understanding of the meaning of war in life. Generals might be expected to commit suicide if they lost a battle and warriors might have to be instantly dispatched if they returned home without their spears, as was the custom among the Matabele, but the governments for which they had fought were rarely moved by the calamities of battle to refashion their defensive posture or redesign their fundamental orientation. Not every society was as totally confident, for example, as the Sukuma of present-day Tanzania, who believed that a victorious enemy could not defeat the spirits of the conquered group or alter their enduring influence on the land, no matter how great the devastation or loss. Yet all accepted with equanimity the ebb and flow of endless war.

Today, Africans in all walks of life continue to be guided by many of these traditional values and institutions, even as they affirm new interests and commitments associated with the lifestyle of the modern age. Intellectually persuasive syntheses of the traditional and the new orders are still rare in African politics. In fact, scholarly analyses of events throughout black Africa (that is, coups d'etat, mutinies, guerrilla operations, revolutions and civil wars) suggest that the two frames of reference may not be easily reconcilable. As Aristide Zolberg rightly notes, "values, norms and structures have survived to a significant extent everywhere, even where their existence was not legally recognized during the colonial era."[21]

---

21 Aristide R. Zolberg, "The Structure of Political Conflict in the New States of Tropical Africa," *American Political Science Review*, March 1968, p. 70 ff. My own analysis of African

The Middle East

Twentieth-century Jews and Arabs are probably more closely tied to traditional religious beliefs than most other literate peoples. Furthermore, their holy texts are different from other sacred literature in an important way: they are not merely intended to be depositories of religious truth, but also serve as comprehensive manuals of instruction in all secular matters. In other words, they are primary and definitive value-references and major resources of normative thinking and policymaking for their respective communities; and in this general context, one cannot read the Old Testament or the Koran and its attendant Islamic traditions without being overwhelmed by the prominence given to the subject of war. According to the Old Testament, which is accepted by the faithful not only as the official history of the Jews but also as a timeless sanction or constitution for the establishment of a Jewish state, there is only one Chosen People; all others are subservient outcasts, subject if necessary to extermination. In Isaiah, chapter 60, the tribal deity advises (in its most benign mood) that "the sons of strangers shall build up thy walls and their kings shall minister unto thee." But elsewhere (Deuteronomy 7, 12, 20; Joshua 1-3, 6, 8; Judges 21; II Kings 3; Psalm 135; and Isaiah 61), we find injunction after injunction on how best to cast out, smite, utterly destroy and extirpate the many "others," great and small, especially those in the region adjoining the River Jordan.

In all the literature exhorting and ennobling war, nothing comes to mind that is quite so chilling as these passages from Deuteronomy, chapter 7:

> 5. But thus shall ye deal with them: ye shall destroy their altars, and break down their images, and cut down their groves, and burn their graven images with fire;
> 6. For thou art a holy people unto the LORD thy God: the LORD thy God hath chosen thee to be a special people unto himself, above all people that are upon the face of the earth.
>
> 16. And thou shalt consume all the people which the LORD thy God shall deliver thee; thine eye shall have no pity upon them; neither shalt thou serve their gods; for that will be a snare unto thee.
>
> 22. And the LORD thy God will put out those nations before thee by little and little; thou mayest not consume them at once, lest the beasts of the field increase upon thee.
> 23. But the LORD thy God shall deliver them unto thee, and shall de-

---

orientations to war and conflict is set forth in *Conflict in Africa: Concepts and Realities* (Princeton, N.J.: Princeton University Press, 1976).

stroy them with a mighty destruction, until they be destroyed.

24. And he shall deliver their kings into thine hand, and thou shalt destroy their name from under heaven: there shall no man be able to stand before thee, until thou hast destroyed them.

25. The graven images of their gods shall ye burn with fire; thou shalt not desire the silver or gold that is on them, nor take it unto thee, lest thou be snared therein: for it is an abomination to the LORD thy God.

26. Neither shalt thou bring an abomination into thine house, lest thou be a cursed thing like it: but thou shalt utterly detest it, and thou shalt utterly abhor it; for it is a cursed thing.

These guidelines for methodical genocide are repeated in Deuteronomy (chapters 12 and 20). With regard to the total destruction of cities delivered by God's will and sword to his people, we read: "Thou shalt save alive nothing that breatheth: But thou shalt utterly destroy them, namely, the Hittites, and the Amorites, the Canaanites, and the Perizzites, the Hivites, and the Jebusites; as the LORD thy God hath commanded thee" (Deut. 20:17). The same divinely sanctioned policy is given expression in II Kings, chapters 22 and 23, where King Josiah is told to break down the enemy, reduce the land to desolation, defile the sepulchers and impoverish all who refuse to acquiesce in the rule of the Chosen Race, and again in Judges 21, where the Chosen are ordered to smite all, including women and children, who do not join them.

Traditional Jewish attitudes toward war and its pursuit must be seen in the context of Near Eastern culture as a whole: throughout the long centuries of ancient history, few if any nations differed from the Jews in their ideas about the conduct of international relations. In other words, war, enslavement and imperialism, unmitigated by considerations of "collective security," "peaceful coexistence" or the "balance of power," combined to make up the real as well as the ideal or preferred system.

In the vast Arab/Islamic domain of West Asia and North Africa, war was idealized and institutionalized in many forms, notably in the *jihad*, or "holy war." Defined in one *hadith* (tradition) as the "peak of religion," the *jihad* is part and parcel of Koranic sacred law. In particular, it denotes the mandate incumbent on each believer to prepare his way to paradise by exerting all his power, including that of the sword, in the service of Allah and the Islamic creed, which is universalist in contrast to the ethnocentric Judaic faith. Consequently, one may view a Muslim's entire life as "a continuous process of warfare, psychological and political, if not strictly military," and conclude that Islamic precepts advance a doctrine of permanent war regardless of whether or not believers are actually engaged in military activities.[22] And, in fact, as the power of the Arabized and Islamized states declined, this doctrine became largely dormant, leaving Muslims in a condition roughly comparable to what is known in international law as a "state of insurgency."

In the context of normative thought, value orientation and foreign-policymaking, then, war is a dominant motif in this culture. Peace, by contrast, being associated with essentially otherworldly, metaphysical concerns, has no overriding positive meaning in temporal affairs, except perhaps as a description of that time when the world will have become Islamized. Pending this outcome of the historic struggle, mankind is divided into the Realm of Peace, whose denizens are engaged in rightful combat at the service of Allah, and the Realm of War, which is the abode, by definition, of all unbelievers regardless of their actual conduct or intentions. It follows logically that diplomacy is viewed more readily as an auxiliary to war, a device serving the cause of belligerence and expansion, rather than an avenue leading toward peace.

Islamic theory grew out of and confirmed the lifestyle of the Bedouin nomads, as shown so convincingly by the biography of the Prophet, the Koran itself, and Charles Doughty's masterful *Travels in Arabia Deserta*. The dominant masculine image or heroic ideal in this harsh world was the warrior, engaged in both great and petty ventures. Camel raids, brigandage, attacks on the despised world of the sedentary and the sown, tribal wars, far-flung military expeditions and, above all, endless wanderings in a hostile environment – all this epitomized the allure and excitement of life that was to compensate for the stark and tedious task of eking out a livelihood. What could peace on earth mean here except sheer boredom, sterility and stagnation?

The political history of the Arabic-speaking peoples from the seventh century to the present corroborates the value system that inhabits their lifestyle and doctrine. Vast expanses of the *dar al-harb* (Realm of War) – in Europe, Africa and Asia – were conquered by force of arms to become integral parts of the Islamic Realm of Peace. Furthermore, Islamic administrations, civil and military, reinforced and perfected their own understanding of the function of diplomacy, borrowing heavily from the sophisticated "warrior diplomacy" of the Persians and Byzantines. This type of statecraft relied on psychological warfare, espionage and subversion in its relentless pursuit of victory over neighboring lands and rulers. In short, nowhere in this region was "peace" accepted as a realizable goal in the conduct of international relations.

---

22   Majid Khadduri, *War and Peace in the Law of Islam* (Baltimore, Md.: The Johns Hopkins University Press, 1955); pp. 55 ff., 62 ff., 144 ff.

The inner order of the Realm of Peace, meanwhile, has also been rent by continual violence and war, even though the ruling idea-system calls for, indeed assumes, peace and unity. The major source of this incongruity has been, and continues to be, the absence of effective fundamental principles of political organization. The caliphate, vaguely conceived by the Prophet's successors, notably the learned ulema (that is, the scholarly divines trained in Muslim law), as the exclusive, indivisible administrative scheme for the governance of the entire community of believers, actually never got off the ground. Instead, commensurate with the swift extension of the faith and culture, we have had multiple caliphates, sultanates and emirates, competitive dynasties, ambitious and contentious aspirants to power, plots and counterplots, assassinations and revolts.

The establishment by conquest of the Ottoman Caliphate in 1453 brought a respite in the divisiveness and anarchy, but its dissolution in 1918-1919 has returned the Arabized Near Eastern Muslims (Turks and Persians can draw from cultural reserves in political organization that are not at the disposal of the Arabs) to more familiar patterns of political thought and action. Contemporary possessors of executive power are thus always tempted to foment or condone violence and intrigue in inter-Arab relations in order to protect their tenuous personal positions or promote their particular dreams of a unity to come.

India

India has experienced the impact of the Middle East (as have parts of Southeast Asia) in a variety of ways, most poignantly perhaps in the fields of statecraft and international relations. Northern India, after all, had been a satrapy of the Persian Empire, and even more extensive portions of the subcontinent were ruled for many centuries by Persianized Mongols. In addition, Islam penetrated through diverse channels to find political expression in the Sultanate of Delhi, the Mogul Empire and, more recently, the Islamic republics of Pakistan and Bangladesh. And yet, many prominent members of the Anglicized elite continue to insist in their scholarly discussions of India's political system that the pre-Islamic Hindu order is still the principal influence, despite massive borrowings – first from the Near East; then, in modern times, from Anglo-Saxon Europe. One of the most ardent Indian nationalists, the late diplomat and historian K. M. Panikkar, thus never tired of reminding his contemporaries in the East and the West that "the society described in the

Mahabharata is not essentially different from what holds its sway today in India," and that if the "Indian administration of today is analysed to its bases, the doctrines and practices of Chanakya [or Kautilya] will be found to be still in force."[23]

Kautilya's Arthasastra, to which Panikkar refers, has been acclaimed as the greatest piece of literature surviving from the Maurya dynasty (322-185 B.C.?). Although other treatises in the same genre exist, the Mauryan chancellor's text is considered exemplary because it explains in systematic fashion how Hindus must think and behave when they are engaged in government, economics and foreign relations. In all these activities, summarily described as the domain of artha (defined by Kautilya as that science which treats of the means of acquiring and maintaining the earth), winning is all that counts. Artha norms are thus carefully set apart in Hindu logic and metaphysics from the codes of conduct mandatory in the pursuit of the three other major ends of life: namely, *kama* (pleasure), *dharma* (duty, especially as it relates to caste regulations) and *moksa* (the assiduous quest for release from life and its illusions). In government and foreign relations, however, the precepts of artha are inextricably enmeshed with the dharma obligations of the warrior caste, for this caste supplies the kings and other secular officers of state, including the armed forces.

The fundamental question — how should men be governed? — was answered in traditional Indian thought and practice by unqualified recourse to *danda*, the rod of punishment. According to the theory of coercive state authority, the king must wield *danda* if he is to enjoy prosperity and acquire not only this world but also the one to come. The Dharmasastras, or Books of the Law, notably the remarkable compendium assigned to Manu;[24] the Arthasastras; the Mahabharata (India's great national epic); and the popular, didactic beast-fables (the best known collection of these being the Pancatantra) thus converge on the doctrine of *matsyanyaya* — the Principle or Law of the Fishes — in accordance with which the king must enforce his government and punish those who deserve it, lest the strong torment the weak as fish are fried on a pike or as in water they devour each other. In deference to the same pessimistic view of human nature, the royal administration relied on espionage as its major agency. As explained in the Mahabharata, a kingdom has its roots in

23 K.M. Panikkar, *A Survey of Indian History* (Bombay: Asia Publishing House, 1954), pp. 2, 29.

24 The Books of the Law and the later didactic portions of the Mahabharata may be taken to represent the post-Mauryan Brahmanic renascence. Muller dates this code later than the fourth century A.D.; Bilhler places it in the second century.

spies and secret agents; therefore, as the wind moves everywhere and pen-
etrates all created beings, so should the king penetrate everywhere by sending
his spies to report disloyalty among subjects, ministers and heirs.

Danda, then, rules all; *danda* is awake while others are asleep; and
danda insists that warriors fight to acquire spiritual merit. These truths are
relayed by all the sacred texts (which continue to be widely read), but most
eloquently by Krishna's discourse with Arjuna in the Bhagavadgita section
of the Mahabharata. The exchange takes place immediately before the great
battle at Kurukshetra. We read that Arjuna, on reaching the battlefield, was
so distressed at the thought of having to fight and kill revered members of his
family, whom he saw ranged on the opposite side, that he resolved to forsake
war. Krishna then turned him from this resolution by reminding him of the
inexorable law of his caste: a *ksatriya* (member of the warrior caste) must fight
and kill his enemy, and the attainment of victory requires total concentration
on the task at hand, including total disregard of other moral or emotional re-
straints.

The same moral teachings have been passed on through the centuries
by other sages and authorities on *artha* and *rajadharma* (royal duties). The king
is created to commit cruel acts, we learn from Bhishma, legendary guru in
the Mahabharata; whereas ordinary men, not made of such stern stuff, seldom
succeed in worldly affairs. Like a snake that devours creatures living in holes,
the earth swallows up the king who does not fight and the Brahman who does
not go abroad (for study).[25]

The history of interkingdom relations before and after the Muslim
conquests faithfully reflects the dictates of the artha philosophy; its annals
speak of endemic anarchy and warfare. True to the law, inequality was pos-
tulated as the everlasting condition of political existence, power as the only
measure of political worth, and war as the normal activity of the state. On the
authority, again, of the Mahabharata:

> Might is above right; right proceeds from might.... Right is in the hands
> of the strong. . . . Everything is pure that comes from the strong. . . .
> When thou findest thyself in a low state, try to lift thyself up, resorting
> to pious as well as to cruel actions. Before practicing morality wait until
> thou art strong. . . . If men think thee soft, they will despise thee. [Book
> XII 134:5-7, 2-3; 140:38; 141:62; 56:21].

A king or politician who has no power is a conquered king, the Arthasastra

---

25 Upendra N. Ghoshal, *A History of Indian Political Ideas: The Ancient Period and the Period of Transition to the Middle Ages* (London: Oxford University Press, 1966), PP. 188 ff, 235.

tells us, and in such a lamentable state of inferiority he is reduced to peace – defined in the Hindu world as stagnation.

Each king, then, was to chart his course of aggression and withdrawal scientifically and realistically. He was to view his own domain as the center or target in the *mandala* (a design symbolic of the universe) of concentric rings of states. His immediate neighbors were by definition his worst enemies; the kings in the second circle were to be viewed as natural friends. The third ring included his enemy's friends, while the fourth was composed of the friends of his allies, and so forth. The science of *artha* instructed the king to be particularly careful when he measured his distance from the dominant state – that is, the state ruled by the king who had the capacity to fight without allies and who was known therefore as the "neutral" king.

Neither the *mandala* nor the particular positions and relations abstracted from it were ever to be trusted completely, however. An arsenal of intelligence tricks and diplomatic techniques, together with a standing and alert army, were regarded as the best security. *Artha* taught the king how to bribe his ally or enemy by gifts, promises and decorations; how to lull him into a sense of false security through conciliation, negotiation and other forms of appeasement, while systematically preparing a military attack on him. Simultaneously, he was of course expected to sow dissension in the frontier provinces of his enemy, in order to soften resistance when he was ready to stage the final armed invasion. In fact, the skills of intrigue were more highly prized by theoreticians and rulers than was material power, as this passage from the Arthasastra illustrates:

> He who has the eye of knowledge and is acquainted with the science of polity, can with little effort make use of his skill for intrigue and can succeed by means of conciliation, and other strategic means, by spies and chemical appliances in over-reaching even those kings who are possessed of enthusiasm and power.[26]

Under such general headings as "Government Based on Deceit," "The Administration of Subversion" and "The Work of an Invader," no subject is given more detailed and devoted attention than that of espionage and "dirty tricks." Brahmans, widows, individuals trained to pose as cripples, saucemakers, dwarfs, eunuchs, courtesans posing as high-class ladies, and merchants are only some of those listed as potential spies or experts in the art of infiltration and sabotage; and complete instructions are provided for their specialized training

26   B. Shamasastry, editor, *Kautilya's Arthasastra* (Mysore: Sri Raghuveer Press, 1951), Book IX, Chap. 1, p. 368.

and deployment. The most favored and talked about category of personnel in this conspiratorial system is that comprising the "shaven heads," ascetics, monks and holy men who have license to conspire and kill in the holy places, and who know how to foment rebellions among the enemy and create other "annoyances in the rear." In "The Battle of Intrigue," it is interesting to note, the Arthasastra makes special mention of "fiery spies" – Asia's earliest guerrillas who conceal "weapons, fire and poison" under their various disguises. Trained "to take advantage of peace and friendship with the enemy," they were charged with the elimination of supply stores, granaries and commanders-in-chief.

To summarize, war was the normal state of affairs in Indian interstate relations until the British unified and pacified the subcontinent. But this episode of Occidental imperialism was a mere moment in the Asian reckoning of time. Hence, many thoughtful and knowledgeable Indians express doubt whether the alien rule of law, including international law, will or should prevail over traditional law.

Southeast Asia

In the course of the fascinating process of cultural diffusion known as the "Indianization" of Asia, the principles of *artha* penetrated much of Southeast Asia. One might even characterize the phenomenon as "cultural imperialism," at least if one were to adopt the parlance currently used by some to describe the impact of European/American culture on the rest of the world. This vast region, which now encompasses Burma, Laos, Cambodia, Thailand, North and South Vietnam, Indonesia, Malaysia, Singapore and the Philippines, was previously dotted with separate kingdoms, each remarkable in its commitment to deeply rooted indigenous beliefs as well as in its talent for integrating appropriate motifs from Hindu, Buddhist, Confucian or Islamic idea-systems.

In pre-nineteenth-century times, the most important regionally unifying themes were the cults of the *devaraja*, or god-king, who could do no wrong so long as he was successful, and the acceptance of rebellion, subversion, war and the threat of war as a normal part of everyday life. Scholars specializing in Southeast Asian history have pointed out that political identity was nowhere a function of secure frontiers, concrete material power, a unifying legal system, or even legitimate royal succession; rather, it depended on an individual ruler's compliance with the cosmo-magical "constitution" of his realm. What mattered in this context – whether in Java, Cambodia, Laos, Thailand-Siam or Burma was physical possession of the capital, the palace and symbolically significant royal regalia; and these sources of prestige could be rightfully seized by cunning or by such acts of violence as the murder of an incumbent prince.

The traditional coexistence of the principles of divine kingship and insurrection explains why the usurper was entitled to obedience and respect, why the idea of the state was associated in the final analysis with the successful ruling personality, and why these kingdoms were locked for centuries in combat of one type or another. Full-scale wars, limited invasions or guerrilla fighting thus marked relations among the rulers of Thailand, Burma, Laos and Khmer-Cambodia, as well as between those of Java and Sumatra. Kingdoms rose and fell, and empires crumbled, only to be resurrected later in some other form. Cambodia, for example, was once part of a Vietnamese empire; the Mekong Delta was constantly in contention; Assam was part of an aggressive Burmese state; and the Khmers, probably the most martial of all these warrior-peoples, tirelessly staked out their claim to what is today Burma. While most principalities and empires were racked by domestic rebellion and subversion, some (notably in present-day Indonesia) are reputed, in modern nationalist texts, to have had vassals as far afield as Vietnam, Cambodia, Thailand and Malaysia. Other allegedly unified kingdoms – Laos and Burma, in particular – were actually conglomerates of separate, warring states.

In this region, then, as in the Middle East and Africa south of the Sahara, internal war merged with external war to form intricate webs of conflict and violence. Hallowed by myth, sanctioned by religion, accepted by the people, and celebrated in legend, art and architecture, this theme has been oft repeated in recent history – in the 1933 "Royalist Rebellion" of Siam, the reinstallation in Burma of the traditional Buddhist trappings of power politics, the elaborate staging by Indonesia's Sukarno of confrontations with Malaysia and the Philippines, and the complex, ongoing interplay of animosities among Cambodia, Thailand and Vietnam. To be sure, one can point to a few nonintervention agreements; for example, in the twelfth century an accord was concluded between Tonkin and the Indianized state of Champa. (It was conceived, by the way, as a reinsurance device that would permit Champa to capture and destroy with impunity the temple of Angkor.) But here, as elsewhere in southern Asia, enmity remained the norm in interkingdom relations. This was true even when China's persistently aggressive policies could have been checked by the organization of collective security measures; instead, each kingdom usually offered separate, ferocious resistance when Chinese forces interfered too blatantly. In short, peace, as a value, had no place either in the metaphysical order of ideas from which these societies derived their identities or in the intricate, artistic processes of statecraft that issued from the royal palaces.

Two thoughts, in particular, impose themselves as one follows the relentless seesaw movements of attack, victory and defeat that have passed like the forces of nature over this culturally complex area. First, in the context of comparative history and religion, there seems to be no doubt that recourse to warfare and palace revolution was hallowed as an integral principle of the ruling cosmic order. And second, this ancient civilization has indeed been the theater in which numerous rival ideas about war – emanating, above all, from India and China – have clashed throughout recorded time. With respect to modern world politics, meanwhile, it is as irresistible as it is ironic to note that American ideas about war were thoroughly discredited precisely here.[27]

China

The *Pax Sinica* that in recent decades has been descending on Tibet and elsewhere in Central Asia, on the Himalayan region of the Indian subcontinent, and on parts of Southeast Asia is a function both of traditional Chinese statecraft and of Mao Tse-tung's adaptations of Marxism-Leninism-Stalinism to the needs of revolutionary China. Contrary to the view held in some American intellectual circles, there is no gulf of discontinuity[28] between the old and the new China when it comes to the politics of war and peace.

With respect to ancient China, as with India, Westerners have long pleased themselves in imagining a spiritually superior civilization, anchored in Confucianism and Taoism, in which men shunned violence and all things uncouth, if only because their attention was riveted on etiquette, sincerity, civility, humanism and the search for harmony. Just why such exalted views of Oriental society should have become so fixed in the Western mind may well be a question that only ethno-psychiatrists can answer as they become adept

---

27 But see *Public Papers of the Presidents of the United States: John F. Kennedy, 1962* (Washington: GPO, 1963), pp. 453-454, for the text of a speech in which this interaction of different ideas about war was intimated, albeit faintly.

28 See, for example, John K. Fairbank, "China's World Order: The Tradition of Chinese Foreign Relations," *Encounter*, December 1966, pp. 14-20; also, by the same eminent scholar, "Introduction: Varieties of the Chinese Military Experience," in Frank A. Kierman, Jr., and John K. Fairbank, editors, *Chinese Ways in Warfare* (Cambridge, Mass.: Harvard University Press, 1974). In Professor Fairbank's view, all Chinese warfare, traditional and modern, is strictly defensive, quite in contrast to that waged throughout the centuries by the Occidental nations, especially the United States, which – he alleges – is invariably aggressive and expansionist. The other essays in this collection provide contrasting views.

at dealing with the symptoms of pathology in inter-cultural relations. Suffice it to say, "[T]here was never a Taoist State as conceived by Chuang Tsu, nor a Confucian State as conceived by Mencius."[29] Indeed, the source materials – which have long been available in excellent translations – teach something else entirely: namely, that China, whatever its geographic configuration and official ideology, has traditionally depended heavily on a judicious investment of war effort, both at home and abroad.

Ping-ti Ho, an authority on Confucian China, and Lucian W. Pye thus agree that the Chinese state has always derived its ultimate power from the army – a circumstance that has largely predetermined its authoritarian character from the days of empire to the rule of Mao Tse-tung.[30] History also instructs us that dynasties usually came to power through armed force; that revolts – and they were commonplace – were staged and smashed by military means; that the science of besieging walled cities was highly developed even in very early times;[31] and that the conduct of all these military operations was organically linked not only to the perfection of weaponry but also, and more importantly, to such official nonmilitary pursuits as the cultivation of crops, the organization of hydraulic works, and the building of walls – occupations without which war could not have proceeded as successfully as it did.[32]

War and agriculture, in fact, have consistently been viewed in China as two fundamental, mutually dependent occupations, perhaps never more so

---

29   Arthur Waley, *Three Ways of Thought in Ancient China* (London: Allen & Unwin, 1939), p. 248; also see pp. 175, 141.

30   Lucian W. Pye, in the foreword to William W. Whitson with Chen-Hsia Huang, *The Chinese High Command: A History of Communist Military Politics, 1927-71* (New York: Praeger, 1973), p. xiii.

31   Numerous references may be found in Kierman, Jr., and Fairbank; note especially Herbert Franke, "Siege and Defense of Towns in Medieval China," pp. 151 ff., 192, for a summation of the technical, administrative and psychological aspects of siegecraft and for comments on the continuity of military strategy and technology in Chinese history. In the same volume, see Charles O. Hucker, "Hu Tsung-hsien's Campaign Against Hsi) Hai, 1556," pp. 273 ff., 305 ff.

32   Edward L. Dryer notes that the regions of China were adminstered from walled cities, which were centers of government as well as places in which large grain reserves were stored. Administrative control of the land was a precondition for further conquest, yet such control could be gained only by capturing the walled cities. These cities thus became the principal military objectives in wars fought within China. The authority of the government over the peasantry, then, extended downward from the walled cities. "Military Continuities: The PLA and Imperial China," chapter 1 in William W. Whitson, editor, *The Military and Political Power in China in the 1970's* (New York: Praeger, 1972), p. 15.

than in the Epoch of the Warring States (ca. 450-221 B.C.) and in the Maoist Period. Accounts from the earlier period can thus be cited to the effect that successive generations of Chinese were decimated by war with methodical regularity, that breathing spaces were allowed only so that the peasant armies might be replenished after having been cut to pieces, and that the army was made to labor on public projects and "in the countryside" when not campaigning.

All-under-Heaven, which consisted of numerous separate provinces, was thus in total disarray during the Epoch of the Warring States; big states ate up lesser ones as systematically as silkworms eat mulberry leaves. Yet China – her contours forever indeterminate – survived mainly, it appears, because the art of war had here reached a mature form by the beginning of the fourth century B.C. By this time, Samuel Griffith notes, the Chinese possessed weapons not at the disposal of other societies and were absolute masters of offensive and defensive tactics and techniques that would have enabled them to cause Alexander the Great a great deal more trouble than did the Greeks, the Persians or the Indians.[33]

All Chinese schools of thought accepted the idea of war, usually as part of the *fa* dimension of government, which existed to supplement what rule by benevolence (*li*) could not accomplish. As Arthur Waley explains, the duty to punish badly ruled states, or to chastise unruly barbarians on the frontiers of the Middle Kingdom, was emphasized consistently by the Confucians and was acknowledged also by their rivals, the Mohists.[34] Just as the principle of filial piety could rightfully be enforced by the killing or mutilation of offspring who resisted paternal guidance, so might the art of persuasion in the community of unequal states be supplemented by the rod of war. In contrast to the domain of internal and family affairs, however, for which legal codes were periodically promulgated by the imperial keeper of Heaven's Mandate, there was no international law and no court or arbitral commission to indicate which state was "badly ruled," or to compose differences impartially. The principle of the "righteous war" thus usually served as a moral cloak for open acts of aggression, which often occurred after atrocity stories had been spread concerning the society singled out for punishment.

The theoreticians and generals who perfected this side of Chinese statecraft, and who then succeeded in bringing about the first unification of

---

33 Sun Tzu, *The Art of War* (New York: Oxford University Press, 1963); translated and edited by Samuel B. Griffith, p. 38.

34 Waley, pp. 141, 152 ff.

China in 221 B.C., are collectively known as the Legalists or Realists. The essence of their science, discernible as early as the seventh century B.C., but which is seen fully developed in the fourth and third centuries in the writings of Sun Tzu, Lord Shang and Han Fei Tzu, is the uncompromising recognition that war and organization for war are the mainstays of government. "How to get the people to die" is the problem that continually occupies the Realists. According to this school of martial thought, it is a misfortune for a prosperous country not to be at war; for in such a state of peace the country will breed the "Six Maggots." In *The Book of Lord Shang*, the parasites that attack in peacetime are enumerated: "rites and music, odes and history, moral culture and virtue, filial piety and brotherly love, sincerity and faith, chastity and integrity, benevolence and righteousness, criticism of the army and being ashamed of fighting. If there are these . . . things, the ruler is unable to make people farm and fight, and then the state will be so poor that it will be dismembered."[35] Vagabonds and draft dodgers, merchants and artisans who deal in nonessential goods, scholars who spread doctrines at variance with Legalist teachings – these are the "Vermin of the State," we learn from Han Fei Tzu.[36] As such, they must be unmercifully quashed so that the people can be kept in ignorance and awe while the king extends the frontiers of the state.

Han Fei Tzu's and Lord Shang's admonitions that the ruler must make certain everyone within his borders understands warfare, that there can be no private exemptions from military service, and that the people must be concentrated on warfare – were faithfully followed by China's first unifier, Shih Huang Ti. The notorious Burning of the Books in 213 B.C. was thus conceived and executed as "the logical last step in unification," as Derk Bodde puts it,[37] and it may now be seen as the precedent for numerous other "cultural revolutions" in imperial and Maoist China.

From time to time, subsequent generations of Chinese scholars have professed to be shocked by these doctrines, but there has never been an age when the martial classics, especially the works of Sun Tzu and Han Fei Tzu,

35 Yang Kung-sun, *The Book of Lord Shang: A Classic of the Chinese School of Law* (Chicago: University of Chicago Press, 1928); translated with introduction and notes by J. J. L. Duyvendak, p. 256. For the proposition that "an intelligent prince ... strives for uniformity, ... restrains volatile scholars and those of frivolous pursuits and makes them all uniformly into farmers," see p. 194.

36 Han Fei Tzu, *Basic Writings* (New York: Columbia University Press, 1964); translated by Burton Watson, pp. 96-117.

37 Derk Bodde, *China's First Unifier: A Study of the Ch'in Dynasty as Seen in the Life of Li Ssu 280?-208 B.C.* (Hong Kong: Hong Kong University Press, 1967), pp. 11, 80 ff.

have not been read. Not only did imperial edicts in later dynasties prescribe the study of these works for the aspirant to an army commission, but Sun Tzu's *Art of War* alone stimulated more than fifty commentaries and interpretative studies between 1368 and 1628. Western nations were long ignorant of the treatise's existence, and when it did become known in the West, the reception was one of neglectful scorn. Japan, by contrast, took Sun Tzu's work most seriously, as did Russia after the Mongol-Tatars brought it there.

In China proper, *The Art of War* continues to be considered a classic to this very day.[38] Conceptually, Maoist strategic doctrine is closely related to the thought of the great master, and Mao Tse-tung's most elegant maxims and most poetic metaphors – which may be found in *Strategic Problems of China's Revolutionary War*, *On Guerrilla Warfare* and *On the Protracted War* – recall those formulated in *The Art of War*. Indeed, reflection on the continuity of Chinese history and ideas about war lead to the conclusion that the Sinification of Leninism could proceed as swiftly and smoothly as it did primarily because of the pervasiveness of Legalism.

Legalist and Maoist ideas converge; they do not meet by chance. And Maoist elites make use of Legalist references deliberately, not in a casually metaphorical way. The struggle between the Legalists, openly identified today with progressive forces in China's past, and the followers of Confucius, who stand for all that is reactionary and regressive in the country's affairs, is thus mentioned in the most improbable contexts. For example, a lengthy stricture on the seemingly mundane subject of traffic safety (broadcast in September 1974 from Haikow, Hainan Island) begins as follows:

> In order to raise traffic-safety work in Hainan to a new level, it is neces-
> sary first of all to do a good job of criticism of Lin [Piao] and Confucius.
> It is necessary to study Marxism-Leninism-Mao Tse-tung thought seri-
> ously and unfold activities to evaluate [sic] the Legalists and criticize the
> Confucianists.[39]

And nothing in Chinese intellectual history suggests that the Chinese communist digestive system would be overburdened by this governmental linkage of traffic-safety work to criticism of Lin and Confucius. After all, as one of the

---

38   See Franke, p. 192. Also see Griffith's remarks on the diffusion of this work, p. 45 ff. and appendix Ill.

39   Joseph Lelyveld, "In China, It's Politics by Allegory," *New York Times*, September 30, 1974. Further see the laudatory essays on the Legalists in *Selected Articles Criticising Lin Piao and Confucius* (Peking: Foreign Languages Press, 1974). This collection includes a "Publisher's Note" to the effect that Lin Piao, being "an out-and-out devotee of Confucius," opposed the Legalist school and attacked the "First Emperor" of the Ch'in dynasty.

foremost students of Chinese military, political and psychological strategy has pointed out, the present system is only the latest manifestation in more than 2,000 years of Chinese strategic thought – a continuity found nowhere in the West.[40]

Chinese commissars, however faithfully schooled in Leninism-Maoism, can thus be expected to use traditional military philosophy to justify both their worldview and the roles assigned them in warfare and society.[41] The easy congruence of these frames of reference is nowhere more impressively demonstrated than in Mao Tse-tung's own writings. Here, elaborate expositions of communist dialectics and discourses on the tactical and strategic doctrines employed during the revolutionary war mingle freely with allusions to Mao's favorite classical novels, traditional boxing precepts, the rules of the ancient game of *wei-ch'i* (in Japanese, *go*) and, above all, to the writings of Sun Tzu, the most esteemed of the Legalist philosophers of war. A few illustrations must suffice.

The Maoists and the Legalists share a militarist, militant vocabulary – one that conveys the unqualified thesis that organization, whether of the village or the world, is war organization, to be established and maintained by the same tactical and strategic rules that apply to the battlefield. Mao thus writes: "In China the main form of struggle is war and the main form of organisation is the army. Other forms, like organisations and mass struggles are also extremely important ... but they are all for the sake of war."[42] The pervasiveness of this conviction explains the stress consistently placed by Legalist and Maoist alike on the need to create agro-military communes and to maintain rural base areas under strict military control. It also explains the striking concurrence of certain poetic metaphors: "The people are like water, and the army is like fish," Mao writes, and the tactics of Chinese statecraft "constitute the art of swimming in the ocean of war." The challenge, as Mao sees it, is "to drown the

---

40 Scott A. Boorman, *The Protracted Game: A wei-ch'i Interpretation of Maoist Revolutionary Strategy* (New York: Oxford University Press, 1969), p. 182.

41 Whitson with Chen-Hsia Huang, p. 438.

42 Mao Tse-tung, *Selected Works* (Revised edition; London: Lawrence & Wishart, 1958), Vol. II, p. 224. And Mao adds: "Every Communist must grasp the truth: 'Political power grows out of the barrel of a gun.' Our principle is that the Party commands the gun, and the gun will never be allowed to command the Party. ... Some people have ridiculed us as advocates of the 'omnipotence of war'; yes, we are, we are the advocates of the omnipotence of the revolutionary war, which is not bad at all, but is good and is Marxist." *Ibid.*, p. 228.

enemy in the ocean of a people's war . . [to] lure him into the deep,"[43] just as it was when Sun Tzu wrote:

> Now the shape of an army resembles water. Take advantage of the enemy's unpreparedness; attack him when he does not expect it; avoid his strength and strike his emptiness, and like water, none can oppose you.
> . . .
>
> Just as water adapts itself to the conformation of the ground, so in war one must be flexible.[44]

Whoever or wherever the enemy is, says Mao, he must be moved "to help in his own destruction," or, as the party chairman puts it elsewhere, he must contribute to his own encirclement. Eventually, "a worldwide net will be formed from which the fascist monkeys can find no escape."[45] Just as "tunneling operations" – vividly described in Sun Tzu's and Mao's manuals – are designed to undermine the physical foundations of the enemy's military position, so are psychological offensives meant to subvert his moral and intellectual bases. Confuse the enemy's leaders; if possible drive them insane, advised Sun Tzu. Costly battles would then become unnecessary. And among the techniques employed to achieve these ends, none has received so much careful elaboration in Legalist and Maoist strategy as the art of dissimulation, simulation and deception. Indeed, as Scott Boorman notes, this concept of stratagem goes far beyond mere attempts to outwit the enemy: it involves the much more sophisticated task of directly manipulating his perception of reality, particularly the values he attributes to various outcomes of the conflict.[46] Sun Tzu's exhortation to "hit the enemy's mind" has thus traditionally been viewed as the prerequisite of victory.

The Legalist master's axiom that "war is based on deception"[47] has been paraphrased often by Mao Tse-tung, who also advocates the intricate, indirect approaches to successful combat and maneuvering outlined in *The Art of War*. Mao's instructions to guerrillas, for example – that they must be

---

43   *Ibid.*, pp. 158, 180.

44   Sun Tzu, pp. 89, 43; see Griffith's editorial comments ("Sun Tzu and Mao Tse-tung"), pp. 45 ff.

45   Mao Tse-tung, pp. 100, 186 ff. The concept of "encirclement," or *wei-ch'i*, relates to the physical as well as the psychological annihilation of the enemy; see Boorman.

46   Scott A. Boorman, "Deception in Chinese Strategy," in Whitson, editor, *The Military* . . . , pp. 313-314. Boorman also points out that current strategic thinking in the United States places comparatively little stress on stratagem.

47   See Sun Tzu, pp. 66-67, 93-94, 97, 106.

as cautious as virgins and as quick as rabbits, mobile and forever changing in appearance – are prefigured in some of Sun Tzu's verses, notably those dealing with offensive strategy and the use of spies and double agents. Furthermore, Sun Tzu's rule that the enemy must be deceived by "creating shapes" or by concealing one's own shape from him is paralleled in Mao's commitment to the consummate skill of creating "illusions":

> Illusions and inadvertence may deprive one of superiority and the initiative. Hence, deliberately to create illusions for the enemy and then spring surprise attacks upon him is a means ... of achieving superiority and seizing the initiative. What are illusions? 'Even the woods and bushes on Mount Pakung look like enemy troops' – this is an example of illusion. And 'making a noise in the east while attacking the west' is a way of creating illusions for the enemy. ... It is therefore extremely important ... to seal off his information, ... keeping the enemy in the dark ... and thus laying the objective basis for his illusions and inadvertence. We are not Duke Hsiang of Sung and have no use for his stupid scruples about benevolence, righteousness and morality in war. In order to win victory we must try our best to seal the eyes and the ears of the enemy, making him blind and deaf, and to create confusion in the minds of the enemy commanders, driving them distracted.[48]

History suggests that the fundamental ideas in a given civilization are often conveyed better by homo ludens – "man the game-player" – than by "man the theory-maker." In China, the idea of war is eloquently expressed in *wei-ch'i*, the game of strategy favored by Chinese statesmen and literati from the early Han dynasty to modern times.[49] Quite unlike the Occidental game of chess, in which the goal is total victory through the capture of a single figure, *wei-ch'i* involves a protracted attempt to extend control slowly over dispersed territory. Play is diffused, and the similarity between this pastime and Maoist guerrilla warfare is quite obvious. The basic strategy in *wei-ch'i* is encirclement and counter-encirclement – all aimed at setting up spheres of influence

---

48 Mao Tse-tung, pp. 172-175. Also see Samuel B. Griffith, *Peking and People's Wars* (New York: Praeger, 1966), p. 33, for a discussion of Mao's advice to guerrillas that they be as cautious as virgins and as quick as rabbits, and for this general approach to tactics: "When you want to fight us we won't let you and you can't even find us. But when we want to fight you, we make sure you can't get away and we hit you squarely on the chin and wipe you out." Cf. Sun Tzu, p. 140.

49 Boorman, *The Protracted Game* ... p. 6; for historical references to the pervasive influence of this game, and the suggestion that Sun Tzu's theories bear a distinct similarity to *wei-ch'i* dicta, see p. 208 (note 8). Boorman's thesis is significant: Chinese communist policies and *wei-ch'i*, he argues, are products of the same strategic tradition – one without a parallel either in Occidental military tradition or in the Western game of chess – and *wei-ch'i* is an important, if little recognized, model of the Maoist system of insurgency.

within enemy territory in order to undermine the opponent gradually by attacks from within. Maoist tactics of "enclosing" or "forming" territory (in the psychological as well as geographic sense) are thus readily comparable to what counts in *wei-ch'i*. Chairman Mao explains:

> Thus the enemy and ourselves each have imposed two kinds of encirclement on the other, resembling in the main a game of weich'i: campaigns and battles between us and the enemy are comparable to the capturing of each other's pieces, and the enemy's strongholds . . . and our guerrilla base areas . .. are comparable to the blank spaces secured on the board.[50]

In the Maoist theory of insurgency, as in *wei-ch'i*, time is long, the grid is large and warfare is continuous, shifting from one sub-board to the next. In either case, success in combat hinges squarely on abiding by Sun Tzu's rule:

> Know the enemy and know yourself; in a hundred battles you will never be in peril.[51]

Or, if one prefers to be up-to-date, by observing Mao Tse-tung's dictum:

> ... war is nothing supernatural, it is one of the things in the world that follow the determined course of their development; hence, Sun Tzu's law, "know your enemy and know yourself, and you can fight a hundred battles without disaster," is still a scientific truth.[52]

IV

The foregoing reflections on war and the clash of ideas support certain general propositions in the fields of international relations and foreign-policymaking.

(1)   There are different cultures in the world. Consequently, there are different modes of thinking, value systems and forms of political organization.

(2)   Within a given society, norms, normative ideas, and notions about what is normal evolve from a continuous interaction between the ruling value system, on the one hand, and the society's perception of social and political reality, on the other.

(3)   A society is virile and effective if it can count on stable patterns of perception, judgment and action. If, by way of contrast, the interaction

---

50   Mao Tse-tung, pp. 151-152.

51   See Sun Tzu, pp. 84, 129.

52   Mao Tse-tung, p. 171.

between the commitment to certain values and the common perception of reality is seriously disturbed, the normative system becomes unreliable; in such circumstances, the society is apt to be morally confused and politically ineffective.

(4) For any society, success in the conduct of international relations turns on two characteristics: (a) confidence in the norms and values that control the inner order of the society and (b) accurate perception of the world in which the national interest must be defined and furthered. Failure ensues when confidence in the nation's integrity is eroded and when the vision of the international environment becomes defective.

(5) In the multicultural environment of the twentieth century, foreign-policymakers must recognize and analyze multiple, distinct cultures as well as political systems that differ from each other significantly in their modes of rational and normative thought, their value orientations, and their dispositions in foreign affairs.

(6) The fundamental foreign policy-related themes running through the histories of sub-Saharan Africa, the Middle East, India, Southeast Asia and China converge on conflict and divisiveness as norm-engendering realities. The evidence shows, in particular, that peace is neither the dominant value nor the norm in foreign relations and that war, far from being perceived as immoral or abnormal, is viewed positively.

(7) This broad concurrence of non-Western traditions stands in marked contrast to the preferences registered in modern Western societies. It is also at odds with the priorities officially established in the charters of the United Nations and affiliated international organizations. To the extent, then, that the United Nations is supposed to reflect universally valid norms, it is a misrepresentation of reality. And insofar as the United Nations was conceived as a norm-creating agency, it has been unsuccessful, particularly with respect to the incidence of war. What is normal in world politics should in these conditions have been inferred pragmatically from the facts.

The challenge of understanding the multifaceted nature of modern warfare has not been met by the academic and political elites of the United States. This failure in the perception of reality has been aggravated by a widespread acquiescence in essentially irrational trends – the inclinations, namely, to dissociate values from facts, to treat values as if they were norms, and to assume that privately or locally preferred values are also globally valid norms.

These intellectual developments have contributed not only to many recent foreign-policy errors but also to widespread uncertainties about America's role in world affairs. They also suggest that the United States has begun to resemble Don Quixote: like the Knight of the Mournful Countenance, it is fighting windmills and losing its bearings in the real world.

3

Adda Bozeman and a Bygone Tradition in Foreign Affairs Analysis that Must Be Revived

John Lenczowski

The passing of Adda Bozeman is a sad day for the study of International Relations in America. For she represented a tradition that modern schools of international affairs seem incapable of continuing. It was a tradition of studying comparative political culture based on profound knowledge of history, religion and ideology. It was very much a product of European education that was transplanted in America and, in the era before the 1960s, produced only a few progeny indigenously. As a result, America has an acute shortage of scholars and officials with Professor Bozeman's capabilities and perspective.

There were several features that characterized Prof. Bozeman's work that made it distinctive enough to be called a "tradition." The first was its emphasis on the study and appreciation of history – not just recent diplomatic history, but the sweep of the history of Western and other major civilizations – and its relevance to contemporary foreign policy. The second was its emphasis on taking ideas, values, religion and other belief systems seriously. The third was its focus on analyzing foreign political cultures on their own terms and avoiding the pitfalls of "mirror-imaging" – a practice of assuming that foreigners think and act "just like us."

The Bozeman tradition contrasts sharply with the main currents of American foreign policy culture, which, stemming from their academic roots, have been excessively materialistic. This culture, focusing on arms, economics and the diplomacy deriving from them, scarcely directs the strategic attention of senior policymakers toward the variables of values, motivation and intention which underlie the disposition of those material things.

Those American scholars who appreciate these more elusive catego- ries that come from ideas, belief and culture tend to hail from the humanistic disciplines of arts and literature. Unfortunately, such scholars rarely if ever study international relations and statecraft. As a result, their talents can almost never be harnessed for practical purposes of diplomacy. Meanwhile, those who do study foreign relations rarely take the time to familiarize themselves with the souls of other cultures via their epic literature, their religions and other customs.

The U.S. intelligence community, for example, has consistently dis- regarded the importance of having its analysts read the character-revealing literature of foreign cultures.

Because of her study of the power of cultural peculiarity, Prof. Boze- man became one of the main protagonists in the debate over how to conduct the recent American policies concerning human rights and the export of lib- eral democracy. Opposing those who believe in the applicability of democracy to most any culture around the world, she argued the difficulty, riskiness and probable futility of exporting democracy to non-Western cultures which do not share the West's unique traditions that have made democratic self-govern- ment possible: liberty; equality; the rule of law; the concept of the sovereignty of the individual in the realms of both rights and responsibilities; the concepts of contract and private property; the Western idea of the state; and the West- ern, but particularly American, dedication to progress.

Whereas others have argued that a number of non-Western cultures as disparate as Japan and India have adopted democracy, Prof. Bozeman coun- tered that without development of the necessary roots in philosophical as- sumptions and customary behavior, democracy in many of these non-Western cultures will merely assume superficial democratic forms which will end up in any case as temporary in the broader historical context.

Genuine Western-style democracy, she argued, cannot be adopted by cultures that emphasize historical continuity rather than progress, social con- straint over liberty, cultural homogeneity rather than diversity, and rejection of the separation of the individual from the group to which he belongs when it comes to questions of rights and responsibilities.

The perspectives Prof. Bozeman brought to this debate shed essen- tial light on the ongoing debate over America's national purpose in its rela- tions with the world. In the original debate, some of America's Founders proclaimed that it was America's mission to see to it that people everywhere around the world should one day be able to live in a political order that protected the inalienable rights of man. Others responded that while America should be a well wisher of freedom for all, she should be a champion and vindicator only of her own. In this way, America would respect the established traditions and forms of order that may be undone usually at considerable risk of greater tyranny or disorder.

While Prof. Bozeman's analysis of this issue has not put the debate to rest, it has reminded people with a more utopian cast of mind that prudence is the sine qua non in any effort to inject Western democratic values into political cultures of vastly different pedigree.

A critical policy implication of the Bozeman tradition is the need to take seriously the role of public diplomacy and related instruments in the overall scheme of American strategy. This entire sphere of foreign relations, which includes cultural diplomacy, information policy, international exchanges, international broadcasting and even psychological operations during periods of crisis and war, has been ignored by both America's academic and official foreign policy cultures.

The neglect of this spectrum of instruments of perceptions management and mutual understanding can only be explained by a systematic failure to take ideas, values, beliefs and their instruments of transmission seriously. This is ironic at the very moment of the dawn of the information age, in a country not unknown for its skill in its advertising, journalism and knowledge industries. It is all the more ironic, given that some of America's public diplomacy endeavors, such as Radio Free Europe, Radio Liberty and the Voice of America have been among the most decisively successful, yet underappreciated, foreign policy enterprises ever devised in recent times.

A final policy implication of the Bozeman tradition that deserves comment is the need for intelligence analysis and diplomacy to take sufficiently into account how much at variance foreign strategic cultures are from ours and how perilous "mirror-imaging" can prove to be. Such analysis can reveal how alien certain foreign traditions are to the American tradition and how incompatible they may be with American diplomatic ends and means. These include such things as: seeking victory and unilateral advantage rather than compromise; having no respect for international borders or Western notions of international law; corrupt business practices; looking upon acts of generosity or accommodation as signs of weakness; having no compunctions about utilizing methods that may be morally objectionable to Americans; etc.

By the same token, her analysis leads one to conclude that certain American predispositions may either be misunderstood, ignored, exploited or scorned by our foreign interlocutors or antagonists. These include such things as: our egalitarianism; our legalistic mentality, our self-fancied pragmatism and rationalism; our tendency to regard trade and commercial relations as indices of mutual accord and harmony and as an instrument to avoid conflict; our penchant for "diplomatism" and dialogue conducted in a spirit that frequently denies the reality of conflicting and sometimes irreconcilable interests; and so forth.

Fortunately, for serious students of statecraft, the written works of Prof. Bozeman remain. But they must be sought out independently, for there are regrettably few in the academic world who would steer students in their direction.

4

Culture Clash-ification:

A Verse to Huntington's Curse

Frederick S. Tipson

We owe to Samuel Huntington a potent provocation,
A trenchant tract to counteract a clear exaggeration:
The notion that the West has won, its culture now supreme,
His book rejects—and then corrects—as wishful in extreme.
For, he insists, our world consists of cultural formations
Arising (and revising) out of eight great civilizations.

He sets our pulses pounding and our wisdom teeth to gnashing
With come-to-blows scenarios of different cultures clashing.
This is of course a tour de force, but somewhere in the tour,
Huntington has been undone by paradigm-amour.

For in his zeal to wheel and deal in fundamental frameworks,
He misses cues and misconstrues just how the global game works.
What Sam doesn't seem to get, despite the implications,
Is that the game has been reset by telecommunications.
Networks and computing make the difference fundamental,
By skewing and redoing social bonds—and governmental.

Since entity identity is much more problematic,
Crash-courses in world politics should not be so dogmatic.
For cultures have been compromised, foundations have been shaken;
Resistant values often overwhelmed or overtaken.
These bouts are seldom brittle, like tectonic plates colliding,
But mushy in the middle—more like pools of ink eliding.

The economic pressures for survival and performance
Have helped to steer the atmosphere to cultural conformance.
The reach of voice and video is widely so invasive,
They bleach each local culture like a very harsh abrasive.
Each center hones distinctive tones, yet even these evolve,
So neither core is like before, and outer realms dissolve.

And though we tow our sentiments and symbols from the past,
Their meaning has been leaning as their context is recast.
While all the cards are shuffled and the deck's in disarray,
The task is how to reavow the order of the day.

Yet Huntington is skeptical of cultural convergence,
And won't endorse the course of what is widely in emergence:
A market-based and liberal-laced embrace of competition,
Against which local cultures often brace in opposition.

These global/local battles are recycled and reheated
In country after country where these ruptures are repeated:
The worldly cosmopolitans confronting long tradition;
Though neither strand may take command, all culture's in transition.
His model may appear to be well-grounded in the past,
But Huntington has pitched his product much too hard and fast.
As he explores these culture wars, he seems to be inviting
What used to be a weakness of the school of "realist" writing,
Ascribing to a concept, like a culture or a state,
A physical reality which doesn't quite equate.

This realist bard from Harvard Yard distorts important factors,
Converting complex cultures into unitary actors;
What Morgenthau and Wolfers did with power among nations,
Huntington has nearly done with full-scale civilizations.
The West, he thinks, is better off consolidating ranks,
Resisting multicultural promoters and their planks,
For cultural dilution means political exposure,
A kind of mind pollution which detracts from value closure.

On human rights the West retains no copyright or patents;

Reforms of basic norms is not a test among combatants.
And so, despite his histrionic altitude and brilliance,
He overstates the length and strength of cultural resilience.
China doesn't merit a description as Confucian
After what has happened since the Maoist Revolution.
Mao's ideas were "Western"–mainly Marxist and Hegelian,
Confused less with Confucius than with Bernard Shaw's Pygmalion–
And surely Deng (although he's sung of spiritual pollution)
Has made the pitch that getting rich can be its own solution.
It isn't that Confucianism isn't there at all,
But, rather, it has undergone enormous overhaul.

Likewise, growing NATO's writ by moving east its proxy
Only reignites the frights of Russian orthodoxy.
It strikes me as a dangerous form of policy confusion,
Boosting culture clashes through a self-fulfilled conclusion.
His quickness has the slickness of a Disney-style cartoon,
The feeling of a foray, or a major trial balloon.
His book conveys a challenge, like he wants us to refute him,
Daring us, by scaring us, to doubt him or dispute him.
Which is fine for academic-argument-displaying
As long as someone powerful won't act on what he's saying.

Because the final irony of Huntington's portrayal
Is that in other countries he may make his biggest sale.
Politicians prone to pick what's overripe or rotten
May resurrect a culture that is gone but not forgotten,
Building on the current state of cultural confusion
To craft a cult of closure or a culture of exclusion.

We publish at our peril and we magnify the dangers
By lending credibility to cultural estrangers.
I prefer a paradigm intent on integration,
The framework for a future forged by acts of innovation:
Taking expectations from technology and trends
And staking aspirations on the future that impends.

All this implies a vision less fixated on our seams

And giving much more weight to global specialized regimes:
Those critical components of a global public order
Of commonsense consensus, both cross-culture and cross-border.
History's indispensable to shape our understanding,
But it needs to be there at the takeoff, not the landing.

To find our voice and tools of choice in shaping human futures,
We need to nurse that vision not with scalpels, but with sutures.
Huntington as scientist may well deduce his stances,
But Huntington as moralist might just reduce our chances.

5

Hybrid Wars

COL John J. Mccuen, USA Ret.

Those who cannot remember the lessons of the past are condemned to repeat it.
—George Santayana[1]

W e in the West are facing a seemingly new form of war – hybrid war.[2] Although conventional in form, the decisive battles in to-day's hybrid wars are fought not on conventional battlegrounds, but on asymmetric battlegrounds within the conflict zone population, the home front population, and the international community population. Irregular, asymmetric battles fought within these populations ultimately determine success or failure.

Hybrid war appears new in that it requires simultaneous rather than sequential success in these diverse but related "population battlegrounds." Learning from the past, today's enemies exploit these new battlegrounds because the West has not yet learned to fight effectively on them. We still do not fully appreciate the impact and complexity of the nuanced human terrain. One need only read our daily newspaper headlines or listen to TV and radio news about the insurgencies being fought within the populations of Afghanistan and Iraq to understand the validity of the above observations. Insurgencies rage within these conflicts' penetrated and often alienated populations in spite of our having first defeated the enemy's conventional forces. Our population at home usually wearies of the protracted struggles, waged, until recently, with little apparent progress. We are in danger of losing if we fail to understand fully the human terrain in these conflicts, as well as, perhaps, the

---

1   George Santayana, *The Life of Reason* (1905-1906), Volume I, *Reason in Common Sense*.

2   "Seemingly" because all wars are potential hybrid wars. Rarely in history have wars ended purely as what we today like to call "conventional."

even more decisive battlegrounds of public opinion at home and abroad.

In the context of hybrid wars, especially at the population level, outcomes should be approached in terms of success or failure rather than the usual military distinctions of victory or defeat. In this regard, the goal or end state sought should be something like "secure improved normalcy," not "defeat the enemy forces" or "overthrow the enemy regime." The critical point is that to win hybrid wars, we have to succeed on three decisive battlegrounds: the conventional battleground; the conflict zone's indigenous population battleground; and the home front and international community battleground.

Merging three battlegrounds and two wars

In spite of the stark lessons of the past – Indochina, Vietnam, Greece, Somalia, and, most recently, Lebanon – we have not yet learned to succeed on the three combined battlegrounds of hybrid war. Military theorists have started to call those conflicts "hybrid wars" or "hybrid warfare" (to include the Army Chief of Staff when he recently announced publication of the new Field Manual [FM] 3.0, Full Spectrum Operations) but few, unfortunately, have talked substantively about how to fight such wars and achieve enduring success.

Thus, hybrid wars are a combination of symmetric and asymmetric war in which intervening forces conduct traditional military operations against enemy military forces and targets while they must simultaneously – and more decisively – attempt to achieve control of the combat zone's indigenous populations by securing and stabilizing them (stability operations). Hybrid conflicts therefore are full spectrum wars with both physical and conceptual dimensions: the former, a struggle against an armed enemy and the latter, a wider struggle for, control and support of the combat zone's indigenous population, the support of the home fronts of the intervening nations, and the support of the international community. In hybrid war, achieving strategic objectives requires success in all of these diverse conventional and asymmetric battlegrounds.

At all levels in a hybrid war's country of conflict, security establishments, government offices and operations, military sites and forces, essential services, and the economy will likely be either destroyed, damaged, or otherwise disrupted. To secure and stabilize the indigenous population, the intervening forces must immediately rebuild or restore security, essential services, local government, self-defense forces and essential elements of the economy.

Historically, hybrid wars have been won or lost within these areas. They are battlegrounds for legitimacy and support in the eyes of the people.

In Vietnam, after a flawed beginning, we learned from our mistakes and successfully won the battle within the South Vietnamese population – although we ultimately lost the war when massive U.S. home front political pressure forced us to withdraw.

From 1968 to 1973, we taught counterinsurgency in every military service school and college. (I myself directed a course on "Internal Defense and Development" at the U.S. Army War College.) However, after Saigon fell, the cry went up, "No more Vietnams!" We dropped all of our courses on counterinsurgency and nation-building and turned our attention to the Cold War, to conventional defensive operations and to the nuclear threat. Later, our armed forces would enter Afghanistan and Iraq without the benefit of a focus on asymmetric war, and with virtually nobody trained in either counterinsurgency or nation-building. After several years of failure in those arenas, we have struggled to relearn, teach, and practice the lost lessons. As our new surge strategy demonstrates, our ultimate success in current conflicts and future interventions will hinge on relearning these lessons of the past.

Fortunately, the Army and Marines have published a new Field Manual (FM) 3-24, Counterinsurgency. But this volume deals primarily with assisting a host nation fighting an insurgency within its population. As noted above, in hybrid wars there likely will be no host government, indigenous military, or police forces at the outset; much of the political, economic, and social infrastructure will also likely have been destroyed or seriously damaged.

These conditions will radically change how our military conducts counterinsurgency. As they have in Afghanistan and Iraq, our military forces will initially have to be responsible for conducting political and economic operations within the population. Until sufficient security and stability have been established to allow other government agencies to first participate and later assume responsibility, military forces will have these burdens while concurrently conducting military operations. FM 3-24 will be valuable as a source for developing hybrid-war strategy, but we will have to use historical lessons and ongoing experience to figure out how to implement strategy.

The Army is also about to issue FM 3-0, Full Spectrum Operations. This manual will be doctrinally vital to the conduct of hybrid war because it acknowledges that we must fight within populations as well as against conventional enemies: "Full spectrum operations are the purposeful, continuous, and simultaneous combinations of offense, defense, and stability...

to dominate the military situation at operational and tactical levels.... They defeat adversaries on land using offensive and defensive operations, and operate with the populace and civil authorities in the area of operations using stability operations."[3] However, like FM 3-24, FM 3-0 will not provide the "how" for operations in either counterinsurgency or hybrid war. Dialogue in professional journals and military schoolhouses, combined with ongoing experience, will help determine the recipes for success in environments peculiar to individual hybrid wars. However, we can no longer delay filling this strategy and doctrine gap – the "how" to fight a hybrid war. That is what I propose we now do.

Use the past to inform the present

Our current enemies have targeted the populations as their battleground of choice. They fully recognize that they do not have the military strength to defeat us in a conventional or nuclear war. However, we have not learned to fight. They know they can protract such wars until home front and international community discouragement over casualties and cost force us to throw in the towel and withdraw. Our enemies' strategic and tactical objectives are thus not to destroy our conventional military forces and seize critical terrain, but to seize, control, and defend critical human terrain until we give up the fight. The decisive battles of the hybrid wars in Iraq and Afghanistan are being fought within the population battlegrounds – the populace in conflict, the home front populations of the intervening nations, and the international community.

As mentioned earlier, one of the most important reasons our enemies have chosen populations as their battlegrounds is that they know that they can protract the war there almost indefinitely. Protraction as a definitive war strategy was first emphasized in modern times by Mao Tse-tung, who promulgated the concept of "protracted revolutionary war." As Mao describes his strategy in his treatise On The Protracted War, "the only way to win ultimate victory lies in a strategically protracted war."[4]

In my 1966-1967 book The Art of Counter-Revolutionary War, I describe Mao's protracted war strategy this way: "To win such a war, the revolutionaries must try to reverse the power relationship '... by wearing down the enemy's

---

3   Field Manual 3-0, Full Spectrum Operations (Washington, DC: U.S. Government Printing Office, 2008).

4   Mao Tse-tung, On the Protracted War, vol. 2, Selected Works (New York: International Publishers, 1954), 180.

strength with the "cumulative effect of many campaigns and battles;" ... by building their own strength through mobilizing the support of the people, establishing bases, and capturing equipment; ... and by gaining outside political and, if possible, military support."[5] These aims reflect the classic essentials of a Fabian insurgency. Our enemies have learned that in hybrid war, protraction wins, especially with its trenchantly modern, technology-enabled impact on spectator populations. Both the insurgent's conventional and information operations are designed to protract the war and gain outside support, thereby wearing down their enemies.

As illustration, the Vietminh insurgents in Indochina knew that the French, using what was essentially an attrition strategy, would lose on both the conventional and population battlegrounds. By fighting a war of attrition, the French could never solve the "mass and disperse" dilemma – whether they should concentrate to defeat the enemy's conventional forces or disperse to protect the population. Nor could they cope successfully with the internationalization of the war, which included enemy safe havens across the Chinese frontier and massive materiel support from the Vietminh's major-power allies, China and (to a lesser extent) the Soviet Union. It was the resulting protraction of the war and its concomitant impact on the French home front that ultimately proved decisive.

Our enemies also know that failing to learn from the French, we initially pursued an attrition strategy in Vietnam. Although we consistently defeated our enemies on the conventional battlefield, we failed to defeat those buried within the population. We judged success by body count – the number of enemy killed or captured – rather than on how many civilians the government protected.

It was our faulty attrition strategy – our failure to orient operations on the population – that deprived the United States of success early in the Vietnam War. Fortunately, General Creighton Abrams assumed command of operations in Vietnam in 1968, and he recognized that the population was the key to success. Abrams radically changed the strategy to embrace a "one-war battlefield" where "clearing, holding, and rebuilding" the population was the critical objective of all military and civilian forces in South Vietnam; in other words, Abrams fought a hybrid war. Significantly, there were adequate U.S.

---

5   John J. McCuen, *The Art of Counter-Revolutionary War – The Strategy of Counter-Insurgency* (Harrisburg, PA: Stackpole Books, 1967), 30. Originally published in 1966 by Faber and Faber in London, *The Art of Counter-Revolutionary War* was republished in 2005 by Hailer Publishing, St Petersburg, FL.

and South Vietnamese forces available, and America was not faced with the "mass and disperse" dilemma. Abrams's new strategy had the salutary secondary effect of greatly reducing the impact of Chinese and Soviet materiel assistance to the insurgency in South Vietnam.

By the end of 1972 and the beginning of 1973, the war within South Vietnam had been essentially won, even in the face of major conventional attacks by the enemy out of their Cambodian and Laotian safe havens. These attacks were defeated with heavy enemy casualties, but, as mentioned earlier, massive internal political pressure on the U.S. government forced it to withdraw in early 1973 under the fiction that South Vietnam would never solve its problems until we withdrew. However, even though the counterinsurgency war within its population had been virtually won, South Vietnam was far from able to defend itself in a conventional war against North Vietnam's 20 battle-hardened, regular divisions without U.S. support. By then, Congress had prohibited this support by law. Also, Washington had never allowed its forces in Vietnam to block the avenues of potential attack within/from Cambodia and Laos. The prohibition against U.S. support and our failure to seal off South Vietnam against future North Vietnamese conventional attacks assured the South's ultimate downfall in the spring of 1975.

This summarized history well illustrates both the genesis of a successful hybrid-war strategy and the potential political pitfalls of fighting hybrid wars. Our ongoing campaigns in Afghanistan and Iraq, as well as the recent Israeli-Hezbollah war in Lebanon, clearly indicate that we in the West – and I include the Israelis in this category – still do not understand how to fight hybrid wars in which the enemy strives to protract war by conducting it within the population while simultaneously attempting to erode confidence at home and abroad as a precursor to military victory. In both Afghanistan and Iraq, our initially victorious conventional attacks created chaotic conditions that allowed stay-behind and outside forces to embed themselves within the population and, seemingly, to protract the war endlessly, thus alienating both the indigenous populations and those at home and abroad. Unfortunately, we created the conditions for protracted war through our own failures, largely by not heeding the lessons of the past.

Fortunately, in Iraq, although belatedly, we have started to use both the lessons of the past and the bitter experience of the present to adopt the hybrid war strategy of securing and stabilizing critical portions of the population and rebuilding the country from the bottom, up. As a result, success has started to replace failure and, though at great cost, we may be finally succeeding or winning in this war.

In a hybrid war, our strategic and operational challenges are the same: how to prevent an enemy from rising up and filling the governmental/services vacuum created behind our advancing forces. If we do not immediately fill this vacuum, we will almost certainly face a protracted insurgent war with the same chaotic features we are now facing in Afghanistan and did face in Iraq.

Of course, we should do extensive planning on how we will establish an indigenous host government, to include military and police forces, and how we will provide protection and essential services to the conflict population. The most critical initial problem in such a campaign will not be how to form a central indigenous government, but how to "clear, hold and build" (our modern doctrine has changed the Abrams-era "rebuild" to "build").

In my book I coined the term "counter-organization" to answer the question of "how." "Counter-organization" calls for us to destroy embedded insurgent organizations and their appeal to the populace by establishing better alternatives. The term is therefore much better suited to the hybrid-war context than anything offered in the new doctrine, which is predicated on having some form of local government and security forces in place. Thus, my proposed hybrid-war strategy would entail a "clear, control, and counter-organize the population" approach.

Counter-organization requires us to seize and maintain the initiative within the population battleground just as surely as we do on the conventional battleground. We must aggressively protect and care for the population. That means we have to "out-guerrilla" the guerrillas and "out-organize" the enemy within the population. We have to carry the war to the enemy by spreading insecurity within his ranks and avoiding it in the population and our own ranks. Clearing, controlling, and counter-organizing the population is the only way to seize the initiative in the human terrain.

In enacting this strategy, the communication of proper values will be critical. Counter-organization necessitates recruiting and training cadres from the local population and then organizing, paying, equipping, and instilling them with values adequate to their task. These values should not necessarily be our values; in fact, they should conform as much as possible to local mores. We must, however, reject the practice of patronage and its attendant corruption so prevalent in many developing societies, ensuring instead that we promote such values as reliability, fairness, and some degree of selflessness in governing, protecting, and supporting the population. We have to maintain or, if necessary, reestablish the fabric of the society the insurgents seek to destabilize.

Building stability by counter-organization, not just from the top down, but, more importantly, from the bottom up, is the way to success. In counter-organization, concentrating efforts and resources at the market-and-village level of the population is essential, since it creates a sense of legitimacy.

The current military term "strategic center of gravity" is the appropriate vehicle for thinking about the elements of success in hybrid wars. Clausewitz defined "center of gravity" as the "hub of all power and movement, on which everything depends... the point at which all our energies should be directed."[6] What all this means for us is that to succeed in a hybrid war, we must first identify proper strategic goals (in military parlance, "strategic end states"), and then go about achieving them by directing all our energies toward accomplishing certain strategic objectives. In hybrid war, we will attain our desired end states only by:

• Conducting conventional operations that carefully take into account how destroying or neutralizing the enemy nation's governmental, political, security, and military structures will play out in the longer term.

• Clearing, controlling, and counter-organizing the indigenous population through a values-oriented approach that fosters legitimacy.

• Winning and maintaining support for the war on the home front(s) and in the international community. Doing so means maintaining legitimacy and avoiding losses through incompetence.

The results of the Indochinese and Vietnam wars and, so far, the results of our present wars in Afghanistan and Iraq, clearly demonstrate that, unless ultimately corrected, failure to succeed in any of these three strategic arenas is likely to result in overall failure.

One battlefield

Considering the expected strategy of virtually all our potential enemies, we will have to be prepared to fight a hybrid war every time we deploy.

It matters not whether the initial conflict begins symmetrically. As evidenced in the world around us and in history, conventional wars are likely to develop major asymmetric components once one force occupies the land of another. At the same time, we will we not be able to dispense with our conventional forces because hybrid war, by definition, will always contain a significant proportion of direct combat by conventional, or even nuclear force. In

---

6 Carl von Clausewitz, *On War*, eds. Michael Howard and Peter Paret (Princeton, NJ: Princeton UP, 1989), 595-96.

addition to developing greater flexibility, we will need to adopt a more holistic attitude to war, approaching the various battlegrounds as one battleground.

Clearly, the conventional aim of defeating the enemy's combat forces has to be achieved at each stage in the campaign. But the decisive second and third objectives, predicated on the populations, must also be achieved.

Not surprisingly, the battle to achieve the third objective of gaining and maintaining public support requires different strategies, tactics, doctrine, and weapons than those used to control the physical and human terrain in combat zones. Competent strategic communications and the perception of moral legitimacy become the determining factors. Our current asymmetric enemies have, with a few exceptions, been much more successful than we have in influencing public perceptions. However, we can reverse this trend and control the moral terrain by judiciously executing our one-war "clear, control, and counter-organize" strategy.

That, of course, does not mean that success among the indigenous population is not decisive – success there is vital to establishing legitimacy and thereby maintaining home and international support. Knowing this, the enemy will often mount combat operations in the field hoping to give the impression that the intervening forces are losing, or at least not winning, and so influence both their enemy's home front and international public opinion.

They expect a country to cut its losses and retreat strategically when public opinion sours. Evidence from Vietnam and Somalia has led them to such a conclusion.

Having adopted a hybrid-war strategy in Afghanistan and Iraq, our adversaries are pursuing goals that remain essentially the same as those of Mao and Fabians in other eras: wear down and wait out the enemy, any way you can. We should not forget that their methods are reminiscent of the way we broke free from Britain in the 18th century – Mao was a great student of our Revolutionary War. Had the British understood counter-organization, things might have turned out very differently here in North America. This enduring commonplace is true, and it flies in the face of assertions from some modern military theorists who try to dissociate themselves from the lessons of the past, arguing that modern wars like those in Afghanistan and Iraq are different in that they are sectarian civil wars. This perspective misses the point.

The point is that all these wars were and are being fought within the indigenous, home front, and international populations at least as much as on the physical battlefields. Understanding that reality means understanding both the threat and the solution. By clearing, controlling, and counter-organizing the population simultaneously with our conventional operations, we can prevent not only insurgencies, but sectarian and civil wars as well. In all such cases, simultaneous achievement of the three strategic aims described above – target lethal force carefully; clear, control, and counter-organize the people; work the information operations – will lead decisively to achievement of strategic objectives and post-bellum success. Thus, from beginning to end, the focus of a campaign must be on aggressively securing and stabilizing the population in the occupied country. Secondarily, there must be constant awareness of the need to maintain the support and assistance of the home front and the world community.

Conclusion

We need to stop planning operationally and strategically as if we were going to be waging two separate wars, one with tanks and guns on a conventional battlefield, the other with security and stabilization of the population. Symmetric and asymmetric operations are critical, interrelated parts of hybrid war, and we must change our military and political culture to perceive, plan, and execute them that way. To become effective modern warriors, we must learn and retain the lessons of the past; we must strategize, plan, and conduct war under a new paradigm – hybrid war.

More than this, we have to change our political and military will. Some worry that after Afghanistan and Iraq we will not have the political and military will to execute, from the start, the sort of costly, complex strategy necessary to succeed in a hybrid war. There is certainly this danger. However, if our vital national interests are threatened, there are likely no good alternatives. Other observers have repeatedly called for negotiating settlements with our adversaries. The problem is that the most dangerous of our enemies, such as Al-Qaeda, summarily reject a negotiated settlement as a violation of their religious law, punishable by death; and most of the others have made endless negotiations, backed by hybrid war – or the threat of it – which they think they can win, a principal pillar of their strategy. Past history and present experience have vividly demonstrated how such strategies have repeatedly eroded our national interests.

Like it or not, political-military will or not, statesmen and soldiers should understand the threat posed by hybrid war. Together, we must develop coherent strategies to avoid or counter such wars, and, if the nation's vital national interests are threatened, we must learn how to fight one successfully.

However, if we do not have the political and military will to fight the hybrid war with the right strategy and resources to support it, we had better not fight it. That is, in itself, a vital national interest.

II
Understanding
the Enemy

6

Avoiding a Napoleonic Ulcer: Bridging the Gap of Cultural Intelligence (Or, Have We Focused on the Wrong Transformation?)

Lt. Col. George W. Smith, Jr., USMC

Not a Frenchman then doubted that such rapid victories must have decided the fate of the Spaniards. We believed, and Europe believed it too, that we had only to march to Madrid to complete the subjection of Spain and to organize the country in the French manner, that is to say, to increase our means of conquest by all the resources of our vanquished enemies. The wars we had hitherto carried on had accustomed us to see in a nation only its military forces and to count for nothing the spirit which animates its citizens.[1]
—Swiss soldier serving in Napoleon's army, 1808

N early two centuries ago, Napoleon Bonaparte preemptively occupied Portugal and Spain and ousted the Spanish royal family for being less than cooperative in supporting his Continental System. As Napoleon proclaimed, "Spaniards, your nation is perishing after a long agony; I have seen your ills, I am about to bring you the remedy for them." Never did he imagine that that conflict would continue in an altogether different form.[2]

The introduction of what was for the first time classified as guerrilla war (or little war, as the Spanish called it) was incomprehensible in Napoleon's conventional military mindset. The resulting resistance, as described by Martin van Creveld, "made do without 'armies,' campaigns, battles, bases, objectives, external and internal lines, *points d'appui*, or even territorial units

---

1 Gabriel H. Lovett, *Napoleon and the Birth of Modern Spain* (New York: New York University Press, 1965), 307.

2 Lawrence Malkin, "The First Spanish Civil War," *Military History Quarterly* (Winter 1989), 20.

clearly separated by a line on a map."[3]

Napoleon's "Spanish ulcer," as he described the Spanish response to his occupation, provides a myriad of timeless lessons for strategic and operational planners. The strategic gap that developed between Napoleon's rapid conventional military victory and the immediate requirement to influence positively the population as part of post-hostilities stabilization operations highlights the limits of conventional military power in post-conflict operations and the perils of forgetting "the people" in the initial and ongoing strategic calculus. Unfortunately, nations and militaries around the globe have been forced to relearn that lesson many times in the ensuing 200 years. The parallels of Napoleon's challenges in Spain with the challenges of contemporary coalition forces in Iraq are striking. While there is a danger in attempting to take historical parallels too far, some similarities are too close to ignore. Moreover, such similarities may reflect the failure to understand the local populace within campaign planning. That understanding forms the bedrock for any successful post-hostility occupation phase.

Thus, cultural intelligence preparation of the battlespace (IPB) with a focus on the post-hostilities landscape is perhaps more important than traditional intelligence preparation of the battlespace, which typically has monopolized the intelligence effort. Countless lessons from history resemble Napoleon's experiences with popular Spanish resistance and provide insight as to what should comprise the proper balance of effort within intelligence preparation for armed intervention. These lessons demonstrate that an inordinate focus on armies at the expense of a focus on the people has and will continue to make winning the peace more difficult than winning the war. Closing the cultural intelligence gap by striking an IPB balance within campaign planning may reduce surprises for an occupying force that historically have impeded the accomplishment of the campaign's stated political or grand strategic objectives.

The Spanish resistance: A historical example

I thought the system easier to change than it has proved in that country, with its corrupt minister, its feeble king, and its shameless, dissolute queen.[4]
—*Napoleon, on the occupation of Spain*

---

3   Martin van Creveld, *The Transformation of War* (New York: The Free Press, 1991), 206.

4   David Gates, *The Spanish Ulcer: A History of the Peninsular War* (New York: W. W. Norton and Company, 1986), 9.

Napoleon gave little thought to the potential challenges of occupying Spain in 1808 once his army had completed what he believed would be little more than a "military promenade."[5] Conditioned by the results and effects of his decisive military victories at Austerlitz (1805) and Jena (1806), Napoleon envisioned that the occupation of major Spanish cities and the awarding of the Spanish throne to his older brother, Joseph, would close the Iberian chapter in his quest for continental domination.

The "ulcer of resistance," which flared up in varying degrees of intensity throughout the country, was most powerful in the territory of Navarre and surrounding northern provinces.[6] That diamond-shaped area, which stretched just under 100 miles from north to south and about 75 miles from east to west, proved to be the hub of Spanish resistance.[7] A closer examination of the inhabitants of that region uncovers numerous clues why resistance to a foreign occupier was so ferocious and weighed heavily in the defeat of Napoleon in Spain. More importantly, it highlights the importance of analysis of the Spanish people, their history, culture, motivations, and potential to support or hinder efforts at achieving French political objectives.

John Tone, in *The Fatal Knot*, succinctly describes the macroconditions for guerrilla resistance in northern Spain: "The English blockade of Spain and Spanish America after 1796 had curtailed the option of emigrating to America, and the economic contraction caused by the blockade made work in Madrid and Ribera more difficult to find as well. What the French found in the Montaña in 1808, therefore, was densely populated, rugged country full of young men with no prospects. Thus, the availability of guerrillas was the result, in part, of a particular economic and demographic conjuncture in the Montaña."[8]

As a whole, the Spanish and Portuguese "were inured to hardship, suspicious of foreigners and well versed in the ways of life – above all, banditry and smuggling – that were characterized by violence and involved constant skirmishes with the security forces."[9] Unknown to Napoleon and his marshals on the heels of another military rout, there bubbled under the surface a

---

5   John Lawrence Tone, *The Fatal Knot* (Chapel Hill: University of North Carolina Press, 1994), 47.

6   *Ibid.*, 6.

7   *Ibid.*, 9.

8   *Ibid.*, 11.

9   Charles J. Esdaile, *The Peninsular War: A New History* (New York: Palgrave McMillan, 2003), 252.

"popular patriotism, religious fanaticism, and an almost hysterical hatred for the French."[10]

The lack of influence of Spanish central authority over its citizenry proved surprising to Napoleon and his marshals, as their point of reference was the occupation of northern European countries. There they found the "Germans and Austrians, conditioned by militarism and centralization, unable or unwilling to act without the permission of their superiors."[11] A common complaint emanating from the French as they grappled with occupying such an independent and spirited Spanish citizenry was that "Spain was at least a century behind the other nations of the continent. The insular situation of the country and the severity of its religious institutions had prevented the Spaniards from taking part in the disputes and controversies which had agitated and enlightened Europe."[12]

Cultural mirror imaging blinded the French to the fact that many Spanish provinces had never been accountable to the royal edicts emanating from Madrid; many Spaniards commonly displayed open contempt for policy disbursed from their national government. Given such an environment of regional independence and domestic political tension, Spaniards even more virulently "disdained anything done for them by a foreigner."[13]

This was especially true in Navarre, where its citizens, imbued with an allegiance to local government and long appeased by national officials in Madrid in an effort to retain a modicum of control, enjoyed perquisites not common in the rest of the country. As Tone wrote: "One of Navarre's most valuable privileges was its separate customs border. In the rest of Spain, the Bourbons had created a single, national market, and they had restricted the importation of finished manufactured goods and the exportation of raw materials in an attempt to encourage industrial development. Navarre, however, controlled its own borders and was exempt from these restrictions."[14]

French preparation of a modicum of cultural intelligence prior to their occupation of Spain might have indicated that the Navarrese stood apart from their countrymen in their relative freedom and therefore would have the

---

10 David G. Chandler, *The Campaigns of Napoleon* (New York: Simon and Schuster, 1966), 659.

11 Tone, 56.

12 Esdaile, 253.

13 Malkin, 20.

14 Tone, 34.

most to lose under French occupation. Succinctly, the Navarrese owed much of their existence to the smuggling of French goods into Spain, avoiding any central government.[15] Cultural analysis might have revealed that assuming new fiscal duties toward an occupying power could be economically ruinous and psychologically offensive to the Navarrese.

The economic factor within the Spanish resistance assumed added significance due to the scattering of Spanish soldiers in the wake of Napoleon's military juggernaut. Dispersed soldiers, no longer sustained by even their paltry military income, were left to roam the countryside focusing simply on survival. According to Charles Esdaile in *The Peninsular War: A New History*: "With the French imposing strict limits on movement and clamping down on many traditional aspects of street life, opportunities to find alternative sources of income were limited, and all the more so as industry was at a standstill and many *señores* [were] unable to pay their existing retainers and domestic servants, let alone take on fresh hands. In short, hunger and despair reigned on all sides."[16]

In such a desperate environment, many young men, former soldiers and civilians alike, were driven into the guerrilla fold out of economic necessity, thus exacerbating the patriotic fervor emanating from northern Spain and further fueled by French occupation.

Napoleon also underestimated the influence of the Catholic Church on the Spanish people. The Church served to energize the notion of an ideological struggle. Ecclesiastical leaders of guerrilla bands were expert at intertwining a host of reasons to continue the struggle against the French. Sébastian Blaze, an officer in Napoleon's army, described the power of the Church: "The monks skillfully employed the influence which they still enjoyed over Spanish credulity ... to inflame the populace and exacerbate the implacable hatred with which they already regarded us. ... In this fashion they encouraged a naturally cruel and barbarous people to commit the most revolting crimes with a clear conscience. They accused us of being Jews, heretics, sorcerers.... As a result, just to be a Frenchman became a crime in the eyes of the country."[17] In the final analysis, "The Spaniards might not have liked their rulers, but they regarded them as preferable to some imposed, foreign dictator. Napoleon could establish Joseph on the throne, but he could

---

15 *Ibid.*

16 Esdaile, 271.

17 *Ibid.*, 252.

not give him popular support."[18]

Napoleon's cultural miscalculation resulted in a protracted struggle of occupation that lasted nearly 6 years and ultimately required approximately three-fifths of the Empire's total armed strength, almost 4 times the force of 80,000 Napoleon originally had designated for this duty.[19] The sapping of the Empire's resources and energy in countering the Spanish resistance had far-reaching implications and proved to be the beginning of the end for Napoleon. He was unfamiliar with this new type of warfare, which was rooted in the people and drove a wedge between conventional military victory and the achievement of his strategic design.

As David Chandler wrote in *The Campaigns of Napoleon*: "Napoleon the statesman had set Napoleon the soldier an impossible task. Consequently, although the immediate military aims were more or less achieved, the long-term requirement of winning popular support for the new regime was hopelessly compromised. The lesson was there for the world to read: military conquest in itself cannot bring about political victory."[20] French grand strategic victory required an understanding as to what winning popular support of the Spanish people actually entailed – a requirement of which Napoleon demonstrated almost complete ignorance. The realities of his tragic oversight were not fully understood until long after conventional combat operations had ceased and various elements of the Spanish population had seized the initiative.

A Preventable "Iraqi Ulcer"?

There is nothing new about the failure to give conflict termination the proper priority. The history of warfare is generally one where the immediate needs of warfighting, tactics, and strategy are given priority over grand strategy. Conflict termination has generally been treated as a secondary priority, and the end of war has often been assumed to lead to a smooth transition to peace or been dealt with in terms of vague plans and ideological hopes.[21]

—Anthony Cordesman

The aftermath of U.S.-led decisive combat operations in Operation Iraqi Freedom has presented challenges to coalition forces similar to those ex-

18   Gates, 11.

19   Chandler, 659.

20   *Ibid.*, 660.

21   Anthony H. Cordesman, "The Lessons of the Iraq War: Issues Relating to Grand Strategy," http://www.csis.org/features/iraqlessons_grandstrategy.pdf.

perienced by the Napoleonic army in Spain almost two centuries ago. Because the harsh treatment of the Spanish citizenry by the French was much different than coalition treatment of the Iraqi people, a parallel cannot be drawn. However, the shared failure to understand the respective peoples and cultures stands in bold relief. The French experience in Spain in 1808, as well as the experiences of many other nations in the intervening 200 years, should drive us to examine why we are prone to making centuries-old mistakes in our campaign planning.

Anthony Zinni, former commander of U.S. Central Command, remarked on the formulation of a coherent campaign design: "We need to talk about not how you win the peace as a separate part of the war, but you have to look at this thing from start to finish. It is not a phased conflict; there is not a fighting part and then another part. It is a nine inning game."[22] In planning for Operation Iraqi Freedom, the coalition was unable to focus its intelligence efforts toward the strategically critical period between the end of large-scale combat and the wholesale transition to stability and support operations until those efforts were too late to be decisive. Planning for post-hostility operations was conducted almost blindly at the tactical and operational levels, with only scattered intelligence on the Iraqi people, what their likely reception of an occupying force might be, and where the coalition might continue to face resistance.

Planners did possess the macro-level detail of the ethnic and religious divisions and the historical tensions between those groups, specifically the Sunnis, Shias, and Kurds. But that cultural understanding did not have the fidelity to highlight, for example, that "more than 75 percent of Iraqis belong to one of 150 tribes, and that significant numbers of Iraqis subscribe to many of the medieval conventions of Islamic law, from unquestioning obedience to tribal elders to polygamy, revenge-killings, and blood money paid to the relatives of persons killed in feuds."[23] Nor did the coalition understand the true depth of influence of the leading Shia cleric, Grand Ayatollah Ali al-Sistani, or the young firebrand, Muqtada al-Sadr.

Furthermore, little analysis was conducted on which segment of the Iraqi population was likely to experience the highest degree of disenfranchisement. Intelligence analysis oriented on the stabilization phase

---

22 Anthony C. Zinni, "Understanding What Victory Is," *Proceedings* 129, no. 10 (October 2003).

23 Patrick Basham and Radwan Masmoudi, "Is President Bush Pushing for Democracy Too Quickly in Post-Saddam Iraq?" *Insight on the News*, December 2003, 46.

failed to account for the prospect of large segments of the Iraqi Republican Guard and Special Republican Guard and remnants of the Ba'athist security apparatus scattered throughout the middle part of the country with no employment and a perceived dim future within an occupied Iraq. In other words, insufficient intelligence focused on the people versus fielded forces and the regime's security apparatus in a post-hostilities scenario.

A broad cultural intelligence analysis, for example, could have drawn out the historical parallel between the Iraqi Sunni Triangle and the Spanish Navarrese Diamond – assuming, of course, that the analysis team was familiar with the cultural factors that contributed to Napoleon's Spanish Ulcer. With that parallel in mind and despite the full benefit of hindsight, few would argue with Anthony Cordesman's assessment in *The Lessons of the Iraq War*: "The Intelligence Community exaggerated the risk of a cohesive Ba'ath resistance in Baghdad, the Sunni Triangle, and Tikrit during the war, and was not prepared to deal with the rise of a much more scattered and marginal resistance by Ba'ath loyalists after the war. The intelligence effort was not capable of distinguishing which towns and areas were likely to be a source of continuing Ba'athist resistance and support."[24]

The U.S.-led planning effort spent more than 16 months determining how best to "break Humpty-Dumpty" with little thought that the coalition might be charged with "putting him back together again." The latter task – infinitely more difficult and foreign to the joint force than tasks associated with conventional combat operations and with the Iraqi people squarely at the center of such a planning challenge – was given short shrift in the intelligence preparation effort. Ironically, tremendous consideration was given to minimizing civilian casualties and collateral damage to critical Iraqi infrastructure needed for follow-on stabilization efforts. However, such analysis and consideration was done largely under the umbrella of "intelligence preparation for combat operations." Moreover, that incomplete analysis failed to recognize the historical truth that the people and the infrastructure bear the brunt of post-combat resistance.

There remained a gap in campaign planning for the period between cessation of major combat operations and wholesale stabilization of the country, a gap that had strategic implications. That historical pitfall is at the root of the following passage from Joint Publication (JP) 5–00.1, Joint Doctrine for Campaign Planning: "Not only must intelligence analysts and planners develop an understanding of the adversary's capabilities and vulnerabilities, they must

24 Cordesman, 24.

take into account the way that friendly forces and actions appear from the adversary's viewpoint. Otherwise, planners may fall into the trap of ascribing to the adversary particular attitudes, values, and reactions that 'mirror-image' U.S. actions in the same situation, or by assuming that the adversary will respond or act in a particular manner."[25]

Much as the French viewed the Spaniards two centuries earlier, U.S. planners were left to peer through an almost exclusively Western lens in their hopeful analysis of how segments of this 25-million-person country might respond to coalition stabilization and support efforts. Succinctly, little professional analysis was conducted to answer the tough questions: "What is it about their society that is so remarkably different in their values, in the way they think, compared to my values and the way I think in my distinctly American way?"[26]

That intelligence gap left too much to wishful thinking and was the context for several broad assumptions that proved to be invalid. Whereas planners left no stone unturned in the intelligence preparation of the battlespace as it related to the defeat of Iraqi forces and ultimate removal of Saddam Hussein, there was little corresponding depth to the analysis of the next target audience within the campaign design, the Iraqi people. Policymakers, commanders, and planners alike were content to lean on the assumption that Iraqis throughout the country would accept the coalition with open arms.

Bridging the gap

We must be cognizant of the changing roles and missions facing the Armed Forces of the United States and ensure that intelligence planning keeps pace with the full range of military operations.[27]

—Hugh Shelton

The U.S. military must accept the fact that the post-hostilities environment is central to campaign design if political objectives are to be achieved. Properly estimating the magnitude of stability and support operations that will be necessary after decisive combat operations end is the only way to prevent

---

25 U.S. Joint Chiefs of Staff, Joint Publication 5–00.1, Joint Doctrine for Campaign Planning (Washington, DC: The Joint Staff, January 25, 2002), II-9–II-10.

26 Joseph L. Strange, *Capital "W" War: A Case for Strategic Principles of War* (Quantico, VA: U.S. Marine Corps University, 1998), 267.

27 U.S. Joint Chiefs of Staff, Joint Publication 2–0, Doctrine for Intelligence Support to Joint Operations (Washington, DC: The Joint Staff, March 9, 2000).

the emergence of a strategic gap. It is the military that will have to grapple with the immediate and diverse challenges that accompany the cessation of large-scale combat operations. More specifically, the military will have to deal with the indigenous population until the arrival of more support-focused and better resourced U.S. agencies and organizations, international aid organizations, and reconstruction specialists.

General Zinni described just such a chaotic environment in an address to the Armed Forces Staff College a decade ago: "The situations you're going to be faced with go far beyond what you're trained for in a very narrow military sense. They become cultural issues; issues of traumatized populations' welfare, food, shelter; issues of government; issues of cultural, ethnic, religious problems; historical issues; economic issues that you have to deal with, that aren't part of the METT–T [mission, enemy, troops, terrain and weather, time available] process, necessarily. And the rigid military thinking can get you in trouble. What you need to know isn't what our intel apparatus is geared to collect for you, and to analyze, and to present to you."[28]

---

28 Strange, 267.

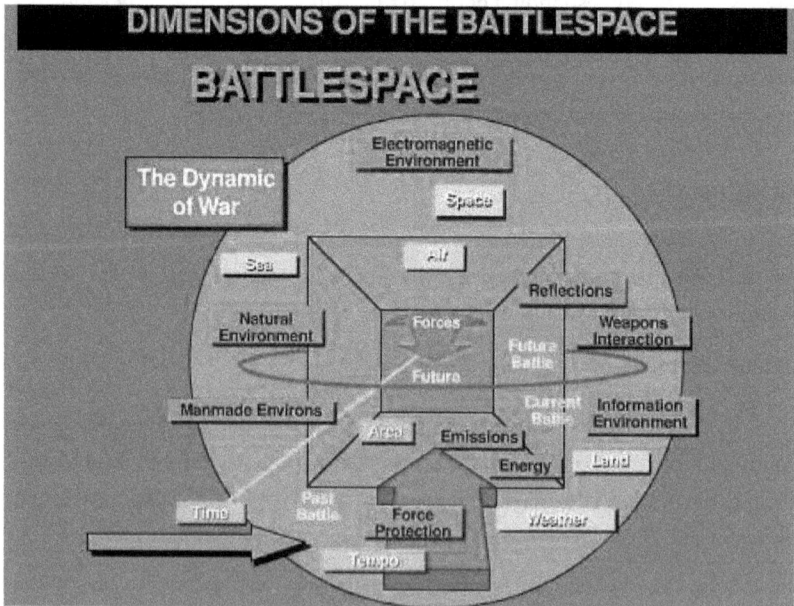

Figure 1. Dimensions of the Battlespace
(Source: Joint Chiefs of Staff, Joint Publication 2–0, Doctrine for Intelligence Support to Joint Operations [Washington, DC: The Joint Staff, March 9, 2000], 1–2.)

While current joint intelligence doctrine that is focused on the people is not barren, the anemic level of detail dedicated to intelligence requirements focused on a people's history and culture is a reflection of the imbalance of the current IPB process. The omission in figure 1 sums up best the mindset of the joint community regarding where "the people" fit within the intelligence requirements for the development of a coherent campaign design.

If properly balanced, a black arrow entitled "People" would be in the center of this diagram opposite the existing black arrow entitled "Forces." This would draw attention to the reality that the civilian population will be the centerpiece of the post-hostilities environment. As currently depicted, this view of the battlespace does little to reinforce the requirements within JP 5–00.1, Joint Doctrine for Campaign Planning, which states that "campaign planners must plan for conflict termination from the outset of the planning process and update these plans as the campaign evolves" and that "emphasizing backward planning, decisionmakers should not take the first step toward hostilities or war without considering the last step."[29]

Furthermore, JP 3–0, Doctrine for Joint Operations, states, "U.S. forces must be dominant in the final stages of an armed conflict by achieving the leverage sufficient to impose a lasting solution."[30] Such leverage toward a lasting solution (grand strategic endstate) can be achieved only if the requisite historical and cultural understanding has been incorporated into the overall planning effort. Currently, joint doctrine for intelligence does not lay the foundation for achieving such leverage. Just as a scan of joint publications suggests that "military professionals embrace the idea of a termination strategy, but doctrine offers little practical help,"[31] a review of doctrine for intelligence preparation of the battlespace reveals only short, topical passages on "The Human Dimension," "The Populace," and the "Effects of the Human Dimension on Military Operations," and only after the various elements of the battlespace contained in figure 1 have been elaborated upon.

Striking a Balance

Our intelligence system is designed to support a Cold War kind of operation. We are

---

29 Joint Doctrine for Campaign Planning, II-4–II-5.

30 U.S. Joint Chiefs of Staff, Joint Publication 3–0, Doctrine for Joint Operations (Washington, DC: The Joint Staff, September 10, 2001), I-10.

31 John R. Boulé II, "Operational Planning and Conflict Termination," Joint Force Quarterly, no. 29 (Autumn-Winter 2001/2002), 97.

"Order of Battle"oriented. We are there to IPB the battlefield.[32]

–Anthony Zinni

The U.S. armed forces must change with that world [a terribly changed and rapidly changing world] and must change in ways that are fundamental—a new human understanding of our environment would be of far more use than any number of brilliant machines. We have fallen in love with the wrong revolution.[33]

–Ralph Peters

With such references to "backward campaign planning" and "achieving leverage," why then do we maintain such an imbalance in our intelligence preparation of the battlespace in the crafting of a holistic campaign design? Or to paraphrase General Zinni, "Why are we only planning for a three-inning ballgame?" One part of the answer may be that Western military forces are not political forces, and professional warfighters like the U.S. and British military tend to see peacemaking and nation building as a diversion from their main mission. It also seems fair to argue that conflict termination and the role of force in ensuring stable peacetime outcomes has always been a weakness in modern military thinking. Tactics and strategy, and military victory, have always had priority over grand strategy and winning the peace.[34]

The gravitational pull of ever-improving technology coupled with the drive toward transformation has compounded the problem by producing a mindset that more can be done with less to achieve the decisive effects in recent and future campaigns. In certain aspects of campaign planning, increased efficiency and effectiveness resulting from technological breakthroughs lend credence to this line of thinking. However, policymakers, commanders, and planners alike must be ever mindful that "efficiency should not be held up as the overarching goal at the expense of better understanding."[35]

Unfortunately, intelligence preparation of the battlespace, the driver of campaign planning, has been co-opted by the same fascination with efficiency. With a heavier focus on the employment of technologically advanced collection systems, the delta between collection efforts focused on enemy forces

---

32 Strange, 266.

33 Ralph Peters, *Fighting for the Future: Will America Triumph?* (Mechanicsburg, PA: Stackpole Books, 1999), 30.

34 Cordesman, 37.

35 Kevin O'Connell and Robert R. Tomes, "Keeping the Information Edge," *Policy Review*, no. 122 (December 2003/January 2004), 19.

and those intelligence efforts focused on the people, "the last six innings of the ballgame" if you will, has actually widened. As Ralph Peters wrote in *Fighting for the Future*, "We need to struggle against our American tendency to focus on hardware and bean counting to attack the more difficult and subtle problems posed by human behavior and regional history."[36]

In the dozen years between Operations Desert Storm and Iraqi Freedom, the U.S. military made tremendous technological strides in its efforts to increase all aspects of its joint warfighting capability, specifically the overall lethality of the force, joint information management, and situational awareness driven by enhanced collection capabilities. But it is clear that the joint force did not place the same premium on gaining an adequate understanding of the Iraqi people and their culture. In analyzing the current situation in Iraq, an astute citizen wrote to the *New York Times*, "There is a crucial need for cultural anthropologists in Iraq even more than capable Arabic speakers. Linguistic knowledge is one thing, but understanding the conventions, subtleties, and nuances of a language and culture is something different."[37]

Three immediate steps should be taken to bridge future cultural intelligence gaps. The first step must be the acceptance that history is important, and while it may not repeat itself as some might argue, it surely holds the clues that will shed light on current and future cultural intelligence requirements. Robert Steele, in *The New Craft of Intelligence*, reinforces the importance of historical analysis: "The first quadrant [requirement], the most fundamental, the most neglected, is that of the lessons of history. When entire volumes are written on anticipating ethnic conflict and history is not mentioned at all, America has indeed become ignorant."[38] Such ignorance would never be tolerated by commanders at any level in preparations for combat operations. That same intolerance must be maintained in planning for missions across the operational spectrum within a comprehensive campaign design.

Yet solving the "puzzle of the people" cannot be the sole domain of military intelligence officials, the small group of foreign or regional area officers, or even the competent but clearly undermanned and overtasked Special Forces, civil affairs, and translator units and detachments sprinkled throughout a large-scale campaign's area of operations. Rather, just as the U.S. defense

---

36   Peters, 45.

37   Mark Rosenman, letter to the editor, The New York Times, October 27, 2003, A–20.

38   Robert D. Steele, "The New Craft of Intelligence: Achieving Asymmetric Advantage in the Face of Nontraditional Threats" (Carlisle, PA: Strategic Studies Institute, 2002), http://www.carlisle.army.mil/ssi/pubs/2002/craft/craft.pdf.

establishment has increased overall efficiency and effectiveness by looking to all corners of the civilian business world within the military hardware acquisition process, so too must the joint force expand its horizons in the development of new intelligence doctrine.

Since doctrine is a guide, the force must be guided in its intelligence activities by those who can shine the strongest beacon on historical and cultural issues. In looking "toward motivational and value similarities, the military should be looking for a few good anthropologists"[39] as well as historians, economists, criminologists, and a host of other experts who can provide the depth of understanding that will lay the foundation for success in post-hostilities operations.

The second step should be a culturally oriented addition to the intelligence series within joint doctrine. The scant references to postconflict intelligence focused on an indigenous population that are currently embedded within several joint publications, namely JP 2–01.3, *Joint Tactics, Techniques, and Procedures for Joint Intelligence Preparation of the Battlespace*, and JP 3–07, *Joint Doctrine for Military Operations Other Than War*, do not adequately address the myriad of unconventional intelligence challenges that are inevitable in the chaos of modern post-hostilities environments. Peters, a career Army intelligence officer, admonishes us: "Military intelligence is perhaps more a prisoner of inherited Cold War structures than is any other branch. ... Our intelligence networks need to regain a tactile human sense and to exploit information technologies without becoming enslaved by them. In most of our recent deployments, no one weapon system, no matter how expensive and technologically mature, has been as valuable as a single culturally competent foreign area officer."[40]

An addition to the intelligence series could take a page or two from the Marine Corps *Small Wars Manual*, which discusses at length the psychology of a country's population. Specifically, it states, "Human reactions cannot be reduced to an exact science, but there are certain principles that should guide our conduct."[41] Furthermore, "These principles are deduced [sic] only by studying the history of the people," and "a study of the racial and social characteristics of the people is made to determine whether to approach them

---

39 Peters, 58.

40 Ibid., 79, 21.

41 U.S. Marine Corps, *Small Wars Manual* (Washington, DC: Government Printing Office, 1940), 18.

directly or indirectly, or employ both means simultaneously."[42]   Finally, the manual warns that "psychological errors may be committed which antagonize the population of a country occupied and all the foreign sympathizers; mistakes may have the most farreaching effect, and it may require a long period to reestablish confidence, respect, and order."[43]

The third step builds on the previous two and bridges the cultural gap through holistic backward planning that achieves intelligence leverage. William Flavin argues for just such a paradigm shift in intelligence preparation of the battlespace in "Planning for Conflict Termination and Post-Conflict Success:" "The IPB should address political, economic, linguistic, religious, demographic, ethnic, psychological, and legal factors. ... The intelligence operation needs to determine the necessary and sufficient conditions that must exist for the conflict to terminate and the post-conflict efforts to succeed."[44]

The U.S. Joint Forces Command, tasked with the lead for transformation within the Department of Defense, has taken a first step in placing more emphasis on cultural intelligence and the imperative to understand a country's or region's dynamics well beyond fielded forces or other potential combatants. The draft "Stability Operation Joint Operating Concept" focuses on the vital period within a campaign that follows large-scale combat operations. As importantly, this concept stresses the requirement for a different focus of intelligence: "Situational understanding requires thorough familiarity with all of the dynamics at work within the joint area of operations: political, economic, social, cultural, religious. The joint stability force commander must have an understanding of who will oppose stabilization efforts and what motivates them to do so."[45]

In reinforcing the fact that the joint force will remain the lead agent for an unspecified period of time upon cessation of hostilities, this concept further highlights the imperative for detailed planning and involvement for a post-hostilities phase across all of the warfighting specialties, specifically intelligence, from the outset of campaign planning. Furthermore, by articulating the critical nature of the period within a campaign when "the joint stability force begins imposing stability throughout the countryside to shape favorable

---

42 Ibid., 28.

43 Ibid., 32.

44 William Flavin, "Planning for Conflict Termination and Post-Conflict Success," *Parameters* 33, no. 3 (Autumn 2003), 101.

45 U.S. Joint Forces Command, draft working paper, "Stability Operations Joint Operating Concept," version 1.03, accessed at http://www.dtic.mil/jointvision/draftstab_joc.doc

conditions in the security environment so that civilian-led activities can begin quickly,"[46] this concept links theater strategic means to grand strategic political endstates. It levies the requirement that intelligence analysis reach depths rarely explored within our current conventional intelligence mindset: "Ongoing human intelligence efforts identify potential cultural, religious, ethnic, racial, political, or economic attitudes that could jeopardize the post-hostility stability operation. The intelligence capabilities begin to focus on the unconventional threat posed by total spoilers. Human intelligence also focuses on the identity, motivation, and intentions of limited and greedy spoilers."[47]

These different categories of spoilers will not be uncovered by conventional intelligence preparation and will remain undetected by our most technologically advanced collection assets. Spoilers will "swim in the sea of the people" and will require a sophisticated and precise intelligence mindset to separate them from the masses and ultimately extinguish the threat they pose to the achievement of the strategic endstate. Such sophistication recognizes that the intelligence focus of the battlespace in post-hostilities must shift from the physical to the cognitive domain, with the paramount concern being the "'minds' of those who might oppose stability."[48]

Conclusion and future implications

What will win the global war on terrorism will be people that can cross the cultural divide. It's an idea often overlooked by people [who] want to build a new firebase or a new national training center for tanks.[49]

–John Abizaid

Proper intelligence preparation of the battlespace focused on the people and the unique challenges of a post-combat operational environment will continue to challenge the joint force in the 21st century, just as it proved to be the Achilles' heel for Napoleon two centuries ago. If we are to apply Napoleon's maxim that "the moral is to the physical as three to one" within a truly holistic campaign design, then perhaps such a ratio should be applied in balancing the collective intelligence effort, with a focus on the people assuming

---

46 Ibid.

47 Ibid.

48 Ibid.

49 Anne Scott Tyson, "A General of Nuance and Candor; Abizaid Brings New Tenor to Mideast Post," *Christian Science Monitor* (March 5, 2004), 1.

paramount importance. That will require addressing intelligence challenges that are unconventional and uncomfortable for planners and commanders at all levels. Comprehensive backward planning with a balanced intelligence effort throughout the breadth and depth of the envisioned campaign will ensure that "forces and assets arrive at the right times and places to support the campaign and that sufficient resources will be available when needed in the later stages of the campaign."[50]

Just as it proved to be the beginning of the end for Napoleon's dominant influence in Europe, giving the importance of "the people" short shrift within the strategic calculus may be the prescription for failure within future military campaigns. Technology is not a panacea within our joint warfighting construct, especially across the spectrum of intelligence requirements. As the world becomes even more complex, it is critical to understand root causes and effects of the histories and cultures of the peoples with whom the joint force will interact. Relying less on high-tech hardware, such a mental shift may be the most transformational step the military can take in preparing for the challenges of the 21st century. These requirements cannot be met with a narrowly focused approach toward intelligence preparation of the battlespace. As Ralph Peters stated at the end of the 20th century: "We will face a dangerous temptation to seek purely technological responses to behavioral challenges—especially given the expense of standing forces. Our cultural strong suit is the ability to balance and integrate the technological with the human, and we must continue to stress getting the balance right."[51]

Sophisticated cultural intelligence preparation of the battlespace may not pinpoint exactly where opposition flashpoints may occur within a post-combat operational environment. However, by achieving appropriate IPB balance, beginning with a bolstered joint intelligence doctrine, the joint force will reduce the potential for strategic gaps by helping to prepare for the Sunni Triangles or Navarrese Diamonds of the future.

If the current modus operandi of insurgents in Iraq is an indicator of the total disregard that future adversaries will have toward global societal norms, the joint force will, in many respects, be operating with one hand tied behind its back. The U.S. military can ill afford to have the other hand bound through the development of comprehensive campaign plans not grounded in solid cultural understanding of countries and regions within which it will likely

---

50 Joint Doctrine for Campaign Planning, II-18.

51 Peters, 14.

operate. To do so risks adding yet another footnote to history highlighting an intelligence gap between combat and stability and support operations.

7

The Military Utility of Understanding Adversary Culture

Montgomery McFate

Cultural knowledge and warfare are inextricably bound. Knowledge of one's adversary as a means to improve military prowess has been sought since Herodotus studied his opponents' conduct during the Persian Wars (490-479 BC). T. E. Lawrence (Lawrence of Arabia) embarked on a similar quest after the 1916 Arab rebellion against the Ottoman Empire, immersing himself deeply in local culture: "Geography, tribal structure, religion, social customs, language, appetites, standards were at my finger-ends. The enemy I knew almost like my own side. I risked myself among them many times, to learn."[1] Since then, countless soldiers have memorized Sun Tzu's dictum: "If you know the enemy and know yourself, you need not fear the result of a hundred battles."

Although "know thy enemy" is one of the first principles of warfare, our military operations and national security decisionmaking have consistently suffered due to lack of knowledge of foreign cultures. As former Secretary of Defense Robert McNamara noted, "I had never visited Indochina, nor did I understand or appreciate its history, language, culture, or values. When it came to Vietnam, we found ourselves setting policy for a region that was terra incognita."[2] Our ethnocentrism, biased assumptions, and mirror-imaging have had negative outcomes. And when people are entering upon a war they do things the wrong way around. Action comes first, and it is only when they have already suffered that they begin to think.

Despite the fact that cultural knowledge has not traditionally been a priority within the Department of Defense (DoD), the ongoing insurgency in

---

1 T.E. Lawrence, quoted in B.H. Liddell Hart, *Lawrence of Arabia* (New York: DeCapo, 1989), 399.

2 Robert S. McNamara, *In Retrospect* (New York: Random House, 1995), 32.

Iraq has served as a wake-up call to the military that adversary culture matters. Soldiers and marines on the ground thoroughly understand that. As
a returning commander from 3rd Infantry Division observed: "I had perfect
situational awareness. What I lacked was cultural awareness. I knew where
every enemy tank was dug in on the outskirts of Tallil. Only problem was, my
soldiers had to fight fanatics charging on foot or in pickups and firing AK-47s
and RPGs [rocket-propelled grenades]. Great technical intelligence. Wrong
enemy."[3] As this commander's observation indicates, understanding one's enemy requires more than a satellite photo of an arms dump. Rather, it requires
an understanding of their interests, habits, intentions, beliefs, social organizations, and political symbols – in other words, their culture.[4]

　　This article argues that new adversaries and operational environments necessitate sharper focus on cultural knowledge of the enemy. A lack
of this knowledge can have grave consequences. Conversely, understanding
adversary culture can make a positive difference strategically, operationally,
and tactically. Although success in future operations will depend on cultural
knowledge, the Department of Defense currently lacks the programs, systems, models, personnel, and organizations to deal with either the existing
threat or the changing environment. A Federal initiative is urgently needed to
incorporate cultural and social knowledge of adversaries into training, education, planning, intelligence, and operations. Across the board, the national
security structure needs to be infused with anthropology, a discipline invented
to support warfighting in the tribal zone.

Changing adversaries and operational environments

　　Cultural knowledge of adversaries should be considered a national
security priority. An immediate transformation in the military conceptual paradigm is necessary for two reasons: first, the nature of the enemy has
changed since the end of the Cold War, and second, the current operational
environment has evolved fundamentally within the past 20 years as a result
of globalization, failed states, and the proliferation of both complex and light
weapons. Although the United States armed and trained for 50 years to defeat

---

3　Steve Israel and Robert Scales, "Iraq Proves It: Military Needs Better Intel," New York Daily
News, January 7, 2004.

4　Culture is "those norms, values, institutions and modes of thinking in a given society that
survive change and remain meaningful to successive generations." Adda Bozeman, Strategic
Intelligence and Statecraft (New York: Brassey's, 1992).

a Cold War adversary, Soviet tanks will never roll through the Fulda Gap. The foe the United States faces today – and is likely to face for years to come – is non-Western in orientation, transnational in scope, non-hierarchical in structure, and clandestine in approach; and it operates outside of the context of the nation-state. Neither Al-Qaeda nor insurgents in Iraq are fighting a Clausewitzian war, where armed conflict is a rational extension of politics by other means. These adversaries neither think nor act like nation-states. Rather, their form of warfare, organizational structure, and motivations are determined by the society and the culture from which they come.

Attacks on coalition troops in the Sunni triangle, for example, follow predictable patterns of tribal warfare: avenging the blood of a relative (*al tha'r*); demonstrating manly courage in battle (*al-muruwwah*); and upholding manly honor (*al-sharaf*).[5] Similarly, Al-Qaeda and its affiliated groups are replicating the Prophet Mohammed's 7th-century process of political consolidation through jihad, including opportunistic use of territories lacking political rulers as a base, formation of a corps of believers as a precursor to mass recruiting, and an evolution in targeting from specific, local targets (such as pagan caravans) to distant powerful adversaries (for instance, the Byzantine Empire). To confront an enemy so deeply moored in history and theology, the U.S. Armed Forces must adopt an ethnographer's view of the world: it is not nation-states but cultures that provide the underlying structures of political life.

Not only our adversaries have changed. The 2001 Quadrennial Defense Review predicted that smaller-scale contingencies – military operations of smaller scale and intensity than major theater or regional wars, such as humanitarian, peacekeeping, peace enforcement, noncombatant evacuation operations, and combating terrorism – will characterize the future operational environment. The use of the military for humanitarian disaster relief, peacekeeping, and counterterrorism operations means that the military will be increasingly forward-deployed in hostile, non-Western environments "disconnected from the global economy."[6] According to Andy Hoehn, former Deputy Assistant Secretary of Defense for Strategy, "The unprecedented destructive power of terrorists – and the recognition that you will have to deal with them before they deal with you – means that we will have to be out acting in the

---

5   Amatzia Baram, "Victory in Iraq, One Tribe at a Time," *The New York Times*, October 28, 2003.

6   Greg Jaffe, "Pentagon Prepares to Scatter Soldiers in Remote Corners," *The Wall Street Journal*, May 27, 2003, 1.

world in places that are very unfamiliar to us. We will have to make them familiar."[7]

Culture matters operationally and strategically

Culture has become something of a DoD buzzword, but does it really matter? The examples below demonstrate three points: misunderstanding culture at a strategic level can produce policies that exacerbate an insurgency; a lack of cultural knowledge at an operational level can lead to negative public opinion; and ignorance of the culture at a tactical level endangers both civilians and troops. There is no doubt that the lack of adversary cultural knowledge can have grave consequences strategically, operationally, and tactically.

At a strategic level, certain policymakers within the Bush administration apparently misunderstood the tribal nature of Iraqi culture and society. They assumed that the civilian apparatus of the government would remain intact after the regime was decapitated by an aerial strike, an internal coup, or a military defeat. In fact, when the United States cut off the hydra's Ba'thist head, power reverted to its most basic and stable form – the tribe. As a tribal leader observed, "We follow the central government. But of course if communications are cut between us and the center, all authority will revert to our sheik."[8] Tribes are the basic organizing social fact of life in Iraq, and the inner circle of the Ba'th Party itself was the purview of one tribe, the Al Bu Nasir. Once the Sunni Ba'thists lost their prestigious jobs, were humiliated in the conflict, and got frozen out through de-Ba'thification, the tribal network became the backbone of the insurgency.[9] The tribal insurgency is a direct result of our misunderstanding the Iraqi culture.

At the operational level, the military misunderstood the system of information transmission in Iraqi society and consequently lost opportunities to influence public opinion. One marine back from Iraq noted, "We were focused on broadcast media and metrics. But this had no impact because the emphasis on force protection prevented Soldiers from visiting coffee shops and buying items on the economy. Soldiers and Marines were unable to establish one-to-one relationships with Iraqis, which are key to both intelligence collection and winning hearts and minds. A related issue is our squelching of Iraqi freedom of speech. Many members of the Coalition Provisional Authority (CPA) and

---

7 Ibid., 1.

8 Baram.

9 Ibid.

Combined Joint Task Force 7 felt that anticoalition and anti-American rhetoric was a threat to security and sought to stop its spread."[10] Closing Muqtada al Sadr's Al Hawza newspaper contributed to an Iraqi perception that Americans do not really support freedom of speech despite their claims to the contrary, reinforcing their view of Americans as hypocrites.

Failure to understand adversary culture can endanger both troops and civilians at a tactical level. Although it may not seem like a priority when bullets are flying, cultural ignorance can kill. Earlier this year, the Office of Naval Research conducted a number of focus groups with Marines returning from Iraq. The Marines were quick to acknowledge their misunderstanding of Iraqi culture, particularly pertaining to physical culture and local symbols, and to point out the consequences of inadequate training. Most alarming were the Iraqis' use of vehement hand gestures, their tendency to move in one's peripheral vision, and their tolerance for physical closeness. One marine noted, "We had to train ourselves that this was not threatening. But we had our fingers on the trigger all the time because they were yelling."[11] A lack of familiarity with local cultural symbols also created problems. For example, in the Western European tradition, a white flag means surrender. Many Marines assumed a black flag was the opposite of surrender – "a big sign that said shoot here!" as one officer pointed out. As a result, many Shia who traditionally fly black flags from their homes as a religious symbol were identified as the enemy and shot at unnecessarily. There were also problems at roadblocks. The American gesture for stop (arm straight, palm out) means welcome in Iraq, while the gesture for go means stop to Iraqis (arm straight, palm down). This and similar misunderstandings have had deadly consequences.

On the other hand, understanding adversary culture can make a positive difference strategically, operationally, and tactically. The examples below illuminate three key points: using preexisting indigenous systems creates legitimacy for the actions of the occupying power, indigenous social organization (including tribal and kinship relationships) determines the structure of the insurgency, and avoiding the imposition of foreign norms will generate public cooperation.

Recognizing and utilizing pre-existing social structures are the key to political stabilization in Iraq. While U.S. policymakers often seemed perplexed by the sub rosa tribal structure in Iraq, the British understood the

---

10 Christopher Varhola, "The U.S. Military in Iraq: Are We Our Own Worst Enemy?" *Practicing Anthropology* 26, no. 4 (2004), 40.

11 John F. Burns, "The Reach of War: The Occupation," *The New York Times*, October 17, 2004.

indigenous system and used it to their advantage. Brigadier Andrew Kennett, commander of the British battlegroup based in Basra, identified a core lesson learned during their history of empire: the importance of adjusting to local cultures and of not imposing alien solutions.[12] In Iraq, the most important element of local culture is the tribe and the associated patronage system. The majority of the population belong to one of the 150 major tribes, the largest containing more than a million members and the smallest a few thousand.

Tribes are invariably patronage systems in which powerful sheiks dispense riches and rewards to sub-sheiks, who in turn distribute resources to the tribal community. Sheiks always need money to generate loyalty from sub-sheiks. There is a saying in Iraq: you cannot buy a tribe, but you can certainly hire one.[13] In Amara, the British did just that. They appointed tribal leaders to local councils and gave the councils large sums to distribute, reinforcing the sheiks' political standing. As one officer noted, "We deal with what exists. In the five months we've been here, we're not going to change the culture of Iraq. We have to work with what there is."[14]

The structure of any insurgency will reflect the indigenous social organization of the geographical region. Thus, charting the Iraqi tribal and kinship system allowed 4th Infantry Division to capture Saddam Hussein. Although most U.S. forces were preoccupied with locating the 55 high-value targets on the Bush administration's list, Major General Raymond Odierno, USA, understood that relationships of blood and tribe were the key to finding Saddam Hussein.[15] Two total novices, Lieutenant Angela Santana and Corporal Harold Engstrom of 104th Military Intelligence Battalion, were assigned to build a chart to help 4th Infantry Division figure out who was hiding Saddam. According to Santana, a former executive secretary, their first thought was "Is he joking? This is impossible. We can't even pronounce these names." Despite the challenges, they created a huge chart called "Mongo Link" depicting key figures with their interrelationships, social status, and last-known locations. Eventually, patterns emerged showing the extensive tribal and family ties to the six main tribes of the Sunni triangle: the Husseins, al-Douris, Hadouthis,

---

12   Neil MacFarquhar, "In Iraq's Tribes, U.S. Faces a Wild Card," *The New York Times*, January 7, 2003.

13   Baram.

14   Charles Clover, "Amid Tribal Feuds, Fear of Ambush and the Traces of the Colonial Past, UK Troops Face up to Basra's Frustrations," *Financial Times* (UK), September 6, 2004, 11.

15   Vernon Loeb, "Clan, Family Ties Called Key to Army's Capture of Hussein," *The Washington Post*, December 16, 2003.

Masliyats, Hassans, and Harimyths, which led directly to Saddam Hussein.[16]

Postconflict reconstruction is most effective when the rebuilt institutions reflect local interests and do not impose external concepts of social organization. For example, Iraqis tend to think of the central government as the enemy. The longstanding disconnect between the center and the periphery meant that Baghdad did not communicate down and city councils could not communicate up. The CPA misunderstood the relationship between Baghdad and the rest of the country and imposed a U.S. model based on central government control. Yet many Marine Corps units intuitively had the right approach and began political development at the local level. A Marine captain was assigned to build a judicial system from the ground up. He refurbished the courthouse, appointed judges, and found the 1950 Iraqi constitution on the Internet. Because he used their system and their law, the Iraqis perceived the court as legitimate. Unfortunately, he was instructed to stop employing Ba'thists. It appears that we are often our own worst enemy.

An inadequate system

Countering insurgency and combating terrorism in the current operational environment demand timely cultural and social knowledge of the adversary. As Andy Marshall, Director of the Office of Net Assessment, has noted, future operations will require an "anthropology-level knowledge of a wide range of cultures." Currently, however, DoD lacks the right programs, systems, models, personnel, and organizations to deal with either the existing threat or the changing environment.

Socio-cultural analysis shops, such as the Strategic Studies Detachment of 4th Psychological Operations Group and the Behavioral Influences Analysis Division of the National Air and Space Intelligence Center, are underfunded, marginalized, and dispersed. Because they lack resources, their information base is often out of date. Task Force 121, for example, was using 19th-century British anthropology to prepare for Afghanistan. With no central resource for cultural analysis, military and policy players who need the information most are left to their own devices. According to a Special Forces colonel assigned to the Under Secretary of Defense for Intelligence, "We literally don't know where to go for information on what makes other societies tick, so we use Google to make policy."

---

16 Farnaz Fassihi, "Charting the Capture of Saddam," *The Wall Street Journal*, December 23, 2003.

Although the Army Intelligence Center at Fort Huachuca, 82nd Airborne Division, Joint Readiness Training Center, Naval Postgraduate School, and John F. Kennedy Special Warfare School all offer some form of predeployment cultural training, their programs are generally rushed, oversimplified, or unavailable to all soldiers and marines who need them. Much so-called cultural awareness training focuses on do's and don'ts and language basics and tends to be geared toward Baghdad.

As one Army colonel noted, "In Western Iraq, it's like it was six centuries ago with the Bedouins in their goat hair tents. It's useless to get cultural briefings on Baghdad." Troops rely on personal reading to make up for the lack of formal training. Inadequate training leads to misperceptions that can complicate operations. For example, those who were instructed that Muslims were highly pious and prayed five times a day lost respect for Iraqis when they found a brewery in Baghdad and men with mistresses. In actuality, Iraq has been a secular society for six decades, and there were relatively few pious Muslims.

Even though all services now have a foreign area officer (FAO) program, the military still lacks advisers who can provide local knowledge to commanders on the ground. The FAO program is intended to develop officers with a combination of regional expertise, political-military awareness, and language qualification to act as a cross-cultural linkage among foreign and U.S. political and military organizations. Because few FAOs are ever subjected to deep cultural immersion totally outside the military structure, most do not develop real cultural and social expertise. Furthermore, most do not work as cultural advisers to commanders on the ground but serve as military attachés, security assistance officers, or instructors. The result is that commanders must fend for themselves. One Marine general explained that his unit had no local experts when it deployed to Afghanistan. The Pastoo-speaking cook on the ship, who happened to be born in Afghanistan, became the "most valuable player" of the mission.

The current intelligence system is also not up to the task of providing the required level of cultural intelligence. Retired Admiral Arthur Cebrowski, USN, Director of the Office of Force Transformation, noted that "the value of military intelligence is exceeded by that of social and cultural intelligence. We need the ability to look, understand, and operate deeply into the fault lines of societies where, increasingly, we find the frontiers of national security."[17]

---

17 Arthur K. Cebrowski, Director of Force Transformation, Office of the Secretary
of Defense, statement before the Subcommittee on Terrorism, Unconventional Treats, and

Rather than a geopolitical perspective, threat analysis must be much more concrete and specific. According to Lieutenant General James Clapper, Jr., USAF, the former director of the Defense Intelligence Agency, "Of course we still provide in-depth orders of battle, targeting data, and traditional military capabilities analysis. But we must also provide the commanders on the ground with detailed information regarding local customs, ethnicity, biographic data, military geography, and infectious diseases." Producing intelligence on these factors can be challenging. As Clapper noted, "We provided detailed analysis on more than 40 clans and subclans operating in Somalia—far more difficult than counting tanks and planes."[18]

Back to the future

A Federal effort is needed to infuse the national security structure with anthropology across the board. While this idea may seem novel, anthropology was developed largely to support the military enterprise. Frequently called "the handmaiden of colonialism," anthropological knowledge contributed to the expansion and consolidation of British power during the era of empire. In the United States, the Department of Defense and its predecessors first recognized culture as a factor in warfare during the Indian Wars of 1865–1885, resulting in the formation of the Bureau of American Ethnology under Major John Wesley Powell. During World War II, anthropologists such as Gregory Bateson served the war effort directly, first conducting intelligence operations in Burma for the Office of Strategic Services, and later advising on how to generate political instability in target countries through a process known as schizmogenesis. American anthropologists produced ethnographies on the Axis powers that facilitated behavioral prediction based on national character.

While Ruth Benedict's 1946 study of Japanese national character, *The Chrysanthemum and the Sword*, is the best known, studies such as Ladislas Farago's *German Psychological Warfare* (1942) collect dust on library shelves. Their predictions were often highly accurate: following recommendations from anthropologists at the Office of War Information, President Franklin Roosevelt

---

Capabilities, Armed Services Committee, United States House of Representatives, February 26, 2004.

18 Lieutenant General James R. Clapper, Jr., "The Worldwide Threat to the United States and Its Interests Abroad," statement to the Senate Committee on Armed Services, January 17, 1995, http://www.totse.com/en/politics/terrorists_and_freedom_fighters/wrldthrt. html.

left the Japanese emperor out of conditions of surrender.[19]

The legacy of World War II anthropology survives in the form of the Human Relation Area Files at Yale University. Established by the Carnegie Foundation, the Office of Naval Research, and the Rockefeller Foundation, this database provided information on Japanese-occupied former German territories of Micronesia. Although the database was maintained for decades after the war with Army, Navy, Air Force, and Central Intelligence Agency funds, U.S. Government agencies seeking "an anthropological-level of knowledge" have sadly now forgotten its existence. During the Vietnam era, the defense community recognized that familiarity with indigenous, non-Western cultures was vital for counterinsurgency operations. The Director of the Defense Department's Advanced Research Projects Agency, R.L. Sproul, testified before Congress in 1965 that "remote area warfare is controlled in a major way by the environment in which the warfare occurs, by the sociological and anthropological characteristics of the people involved in the war, and by the nature of the conflict itself." To win hearts and minds, counterinsurgency forces must understand and employ local culture as part of a larger political solution. As General Sir Gerald Templer explained during the Malayan Emergency, "The answer lies not with putting more boots into the jungle, but in winning the hearts and minds of the Malayan people." Thus, the U.S. defense community determined that it must recruit cultural and social experts. Seymour Deitchman, DoD Special Assistant for Counterinsurgency, explained to a congressional subcommittee in 1965: "The Defense Department has... recognized that part of its research and development efforts to support counterinsurgency operations must be oriented toward the people... involved in this type of war; and the DoD has called on the types of scientists – anthropologists, psychologists, sociologists, political scientists, economists – whose professional orientation to human behavior would enable them to make useful contributions in this area."[20]

During the Vietnam era, the special warfare community understood that success in unconventional warfare depended on understanding indigenous, non-Western societies, and they turned to anthropologists. U.S. Special Operations Command's Special Operations in Peace and War defines unconventional warfare as "military and paramilitary operations conducted by

---

19   David Price, "Lessons from Second World War Anthropology," *Anthropology Today* 18, no. 3 (June 2002), 19.

20   Irving Louis Horowitz, ed., *The Rise and Fall of Project Camelot: Studies in the Relationship Between Social Science and Practical Politics* (Cambridge, MA: MIT Press, 1967).

indigenous or surrogate forces who are organized, trained, equipped, and directed by an external source." To conduct operations "by, with, and through," Special Forces units must have the support of the local population, which can be decidedly difficult to secure. While he was acting as an adviser to U.S. troops in Vietnam in 1965, British expert Sir Robert Thompson suggested that anthropologists be used to recruit aboriginal tribesmen as partisans. Indeed, anthropologists excelled at bridging the gap between the military and tribes. Special Forces in Vietnam, for example, were assisted by Gerald Hickey in working with the Montagnards.

So where are the anthropologists now that the government needs them? Although the discipline's roots are deeply entwined with the military, few anthropologists are interested in national security. Their suspicion of military activity stems from a question of ethics: if professional anthropologists are morally obliged to protect those they study, does their cooperation with military and intelligence operations violate the prime directive? They believe it does. This conclusion was based on a number of defense projects that sought to use anthropological tools in potentially harmful ways. In 1964, the Army launched Project Camelot, a multinational social science research project, to predict and influence politically significant aspects of social change that would either stabilize or destabilize developing countries. The effort was canceled in July 1965 after international protests erupted in target countries. Critics called Camelot an egregious case of "sociological snooping."[21]

While anthropological knowledge is now necessary to national security, the ethics of anthropologists must be taken into account. In addition to direct discussion and debate on using ethnographic information, policymakers and military personnel must be trained to apply anthropological and social knowledge effectively, appropriately, and ethically. The changing nature of warfare requires a deeper understanding of adversary culture. The more unconventional the adversary, and the further from Western cultural norms, the more we need to understand the society and underlying cultural dynamics. To defeat non-Western opponents who are transnational in scope, nonhierarchical in structure, clandestine in approach, and who operate outside the context of nation-states, we need to improve our capacity to understand foreign cultures.

The danger is that we assume that technical solutions are sufficient and that we therefore fail to delve deeply enough into the complexity of other societies. As Robert Tilman pointed out in a seminal article in *Military Review*

21 Ibid., 47–49, 232–236.

in 1966, British counterinsurgency in Malaya succeeded because it took account of tribal and ethnic distinctions, while similar U.S. efforts in Vietnam were bound to fail because they lacked anthropological finesse.[22]

---

22 Robert O. Tilman, "The Nonlessons of the Malayan Emergency," *Military Review* 46 (December 1966), 62.

8

Center of Gravity and Asymmetric Conflict: Factoring In Culture

John W. Jandora

This essay addresses asymmetric conflict in its current manifestation, which has come to be called jihadism. It accepts that the concept of center of gravity is applicable to such conflict, as has been argued by many study projects at the U.S. Army War College.[1] These studies, however, do not extend to the resistance struggle in Iraq. Even in their treatment of Al-Qaeda, they disagree as to what constitutes its center of gravity and reflect questionable assumptions about Islamist militancy. Departing from the conventional systemic approach, the present study focuses on contrast of culture to tie together loose strings and add clarity to the dynamic of jihadism.

To begin with, center of gravity in the context of asymmetry has no correlation with the disposition, maneuverability, or sustainability of a field force or to the capacity of states to mobilize assets of manpower and materiel. Nonetheless, the term remains applicable, particularly as used by Antulio Echevarria. In his treatise on "Clausewitz's Center of Gravity," Echevarria reinterprets Clausewitz's words as advice to look first for unity of effort and then "for connections among the various parts of an adversary, or adversaries, in order to determine what holds them together," as if by centripetal force. "Centers of gravity are focal points that serve to hold a combatant's entire sys-

---

1 The relevant works, all products of the Strategy Research Project at the U.S. Army War College, are James A. Bliss, "Al-Qaeda's Center of Gravity" (Carlisle Barracks, PA: Strategy Research Project, U.S. Army War College, May 3, 2004); Stephen W. Davis, "Center of Gravity and the War on Terrorism" (April 7, 2003); Joe E. Ethridge, Jr., "Center of Gravity Determination in the Global War on Terrorism" (May 3, 2004); John L. Haberkern, "The Global War on Terrorism: Ideology as Its Strategy Center of Gravity" (May 3, 2004); James Reilly, "A Strategic Level Center of Gravity Analysis on the Global War on Terrorism" (April 9, 2002); Joseph P. Schweitzer, "Al-Qaeda: Center of Gravity and Decision Points" (April 7, 2003).

tem or structure together and that draw power from a variety of sources and provide it with purpose and direction."[2]

The term asymmetric warfare similarly deserves clarification. The base concept of a weaker adversary using unconventional means, stratagems, or niche capabilities to overcome a stronger power remains pertinent. However, the original hypotheses of rogue states launching chemical, biological, and radiological attacks or millennialist terrorists wreaking havoc in the United States have been supplanted by the realities of the 9/11 attacks, the Taliban/Al-Qaeda aggression in Afghanistan, and the Sunni resistance in Iraq. The common denominator of these realities is the legitimizing of hostile action through the tenet of Jihad – the Islamic imperative of fighting infidels to regain independence of action on the micro level or to bring social justice and ultimately salvation to mankind on the macro level. Thus, asymmetric conflict has become associated with Jihadism. As any complex word-symbol, Jihad lends itself to various interpretations, including who may rightfully invoke it and how it may be conducted. Such considerations notwithstanding, Jihadism is the hallmark of America's current opponents.

Given this delimitation of asymmetric conflict, jihadism manifests itself to the U.S. military as an array of relatively small-scale, low-level attacks by tribal militias, armed brotherhoods (Sufi militias), factional/party militias, outlaw gangs, and militant cells. This phenomenon is very different from the long-held image of companies or battalions deployed "as two up and one back" – doctrinal, spatially structured combat by state-organized forces. It does not, however, defy analysis of force generation and sustainment. Hence, this essay seeks to expose and explain the centripetal (in-drawing) force that binds the disparate elements in their asymmetric approaches to jihad. The process results in finding centers of gravity.

Tribes and clients

Two countervailing social forces shape the jihadist community, tribalism and clientelism. Both are outside the experience of most Americans. Both terms generally evoke disdain, albeit for quite different reasons.[3] Tribalism,

---

2 Antulio J. Echevarria, "Clausewitz's Center of Gravity: Changing Our Warfighting Doctrine—Again!" (Carlisle Barracks, PA: Strategic Studies Institute, September 2002), vi–vii, www.carlisle.army.mil/usassi/welcome.htm

3 Some of that controversy is noted in Howard Handleman, *The Challenge of Third World Development*, 3d ed. (Saddle River, NJ: Prentice Hall, 2003), 86.

as a derivative of tribe, is problematic because many scholars contend that the base term lacks specificity and therefore analytic usefulness. Clientelism, on the other hand, evokes images of the old-time, party-linked patronage politics of America's big cities, which the school of political correctness sees as deserving avoidance if not censure. Disdain notwithstanding, anyone who has lived beyond the Western enclave in most of the Islamic world knows such terms are indispensable.

The scholarly critique of the term tribe draws attention to its seemingly arbitrary use to denote groups as small as extended families and as large as nationalities. The head count of a tribe correspondingly ranges from a few score to hundreds of thousands. There is controversy whether the term applies to urban as well as rural populations and where the distinction lies between clan and tribe. Moreover, genetic linkage may not correlate with tribal alliances and rivalries. However, none of these objections are critical because tribe and tribalism can indeed be defined in practical terms. According to William R. Polk, a tribe is a kinship group that is optimally sized to its ecologic setting – large enough to accomplish its minimal economic chores and defend itself but small enough to keep members in contact and remain proportional to the supply of food. Thus, history reveals that "clans were constantly splitting apart as they grew beyond their resources or as their resources contracted in times of drought or seasonal change, [and] some of us were periodically becoming them."[4] This process is depicted in the Bible, when the extended families of Isaac and Ishmael, the sons of Abraham, drew apart, evolving into two distinct and now antagonistic peoples, the Jews and Arabs. Then, too, there is evidence that tribes intermingled for ecologic reasons, yet upheld a myth of common ancestry.

If we accept tribe as a valid term of analysis, we can proceed to a meaningful definition of tribalism. It is not the antithesis of globalism, as some scholars suggest, nor a primitive form of nationalism. Rather, it is the self-legitimation of the kin group and its intent and endeavor to optimize its collective self-interest. Self-legitimation is conviction that the tribe is the beginning and end of loyalty, identity, obligation, purpose, status, honor, past, and future – exclusiveness relative to society at large. Thus, the tribe constitutes its own armed force – a militia consisting of most or all fit adult males. The influence of tribalism may be strong or weak, depending on such circumstances as affronts to honor, threats to security, challenges to livelihood, or summons to

---

4 William R. Polk, *Neighbors and Strangers: The Fundamentals of Foreign Affairs* (Chicago: University of Chicago Press, 1997), 30, 218.

jihad. Circumstances may lead to voluntary or compulsory compromises with kin group exclusiveness. (See figure 1 for a depiction of this phenomenon.) Individual tribesmen may be compelled to serve in the state's military establishment or voluntarily join the party that rules the state. At a higher degree of drift, they may voluntarily leave their homeland at the behest of some militant preacher to join a mujahideen group. However, the tribal bond remains unbroken except in cases of full self-alienation. Up to that extreme point, the individual expects, and is expected, to serve tribal interests. He will give the needs of his kinsmen priority and respond to the directives or entreaties of the tribal authority.

It is at the point of full self-alienation that clientelism prevails: individuals stop acting as tribesmen and unquestioningly submit to the authority of preachers or operational leaders. This phenomenon, which involves a small minority, has parallels in Western societies, where youths alienate themselves from their families to follow cult leaders. In both cases, the leader (patron) offers the followers (clients) religious salvation in return for loyal service. The comparison has limits because the personality factor – adulation of the leader – seems more significant in the Western case than in Islam. Osama bin Laden himself seems to be creating a cult of personality through his media releases, but this may be a hasty interpretation. It is noteworthy that his harangues are largely cast in nonegotistical terms, phrasing in grammatical third person (it, that) rather than first person (I, me). Neither his deputies nor the leaders of allied militant groups seem to exploit a personality factor. In their propaganda, the more infamous actors pledge to cooperate with Al-Qaeda and recognize bin Laden as head. However, such allegiance is based on volition, not obligation as is the case with tribalism. Hence, it seems that the militant group leaders attract followers from both self-alienated individuals and genuine outcasts by justifying and facilitating jihad.

Figure 1: Compromise of Kin Group Exclusiveness.

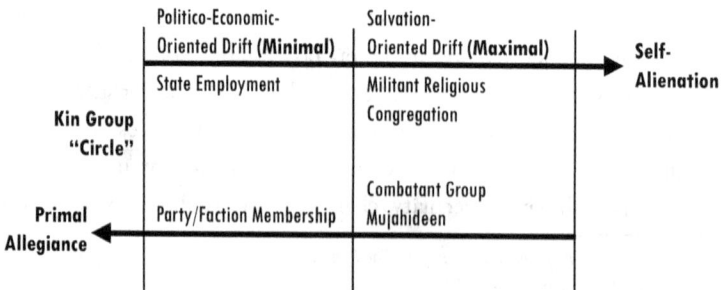

	Politico-Economic-Oriented Drift **(Minimal)**	Salvation-Oriented Drift **(Maximal)**	**Self-Alienation** →
Kin Group "Circle"	State Employment	Militant Religious Congregation	
Primal Allegiance ←	Party/Faction Membership	Combatant Group Mujahideen	

Coopting tribal authority

There is certainly give and take, and even some overlap, between the competing influences of clientelism and tribalism. The Ba'thist resistance in Iraq ostensibly derives motive from old party ideology, yet it must heavily resort to the tribal environment for manpower, subsistence, weapons caching, smuggling assistance, and safe haven since the party/state structure has been destroyed. The real authority within a tribe might be contested among its nominal chief, council of elders, or religious leaders. The outcome might determine whether the males of the tribe mobilize together as an integral tribal militia or component of a Sufi militia, or go off individually to a mujahideen camp.

Such variances should not be daunting, however, because graphing them affords the necessary perspective on the adversary. Figure 2 depicts the resistance construct for Iraq, which includes a small foreign component and a much larger native component. The graphing of the native component indicates tribes with members in the Salafist (religious militant) and Ba'thist arenas as well as the military-security establishment, where they covertly facilitate resistance activity. The tribal leaders have plausible denial insofar as the tribal militias as such are not committed against multinational forces, while the tribe has ostensibly committed assets to the new regime.

In figure 2, space B, at which these various arenas (shown as circles) overlap, is tribal authority, which in this scenario abets the resistance. That space is both physical and moral; it consists of the tribal assemblies where decisions are made as well as the beliefs and rituals that legitimize such decisionmaking. Space B, with its two dimensions, is thus the notional center of gravity for tribally connected resistance. Reversing the scenario perhaps better illustrates the point. Should the tribal authority opt to support the new

Figure 2: Sunni Resistance Construct.

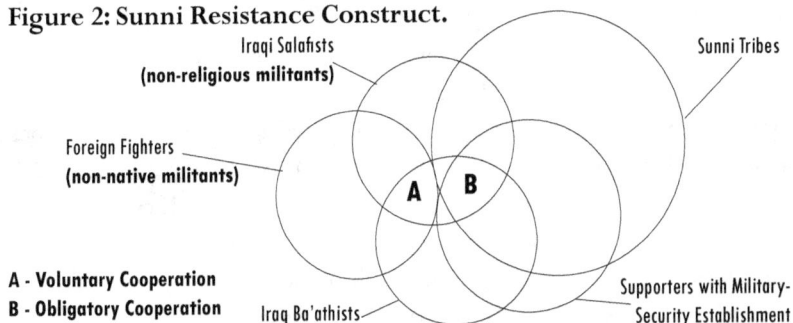

Iraqi Salafists
(non-religious militants)

Sunni Tribes

Foreign Fighters
(non-native militants)

A   B

A - Voluntary Cooperation
B - Obligatory Cooperation

Iraq Ba'athists

Supporters with Military-
Security Establishment

government, the tribal members in the ranks of the military-security forces would cease their subversive activity, and those acting out Ba'thist or radical Salafist agendas would cease hostilities. From either perspective, the associated critical capabilities of tribal authority are ensuring that the kin group has economic sustenance and security from threats.

Specifying a center of gravity, however, is far from devising an effective strategy. The first relevant consideration is that the two dimensions of tribal leadership are not equally approachable. Addressing the moral dimension would be a generational project and is, therefore, a non-starter. On the other hand, addressing the physical dimension is more feasible and suggests two approaches: removing tribal leadership, hopefully without provoking greater antagonism, or coopting tribal authorities and, through them, their tribes. Still, determining the best course of action requires many other considerations. There are hundreds of tribes in Iraq, as there are in Afghanistan and many other Islamic countries. Within a country, some tribes are more powerful than others, some are bitter rivals, and some have regional dominance without ranking very high in the national pecking order.

These power relations can of course be uncovered and then factored into a counterresistance strategy. The operative questions are:
• Which tribes are most significant in terms of manpower, control of strategic terrain or resources, external influence, and historic role?
• Which tribes will resist cooptation, either as a matter of principle or as a matter of irreconcilable rivalry?
• Regarding those tribes open to cooptation, what is the cost of coopting them, for example, in terms of money, official positions, local development projects, or public sector employment? Is the cost bearable?
• What are the tradeoffs between coopting at the regional or subregional versus the country level?
• How does the stabilization force contain the tribes that cannot be coopted, for whatever reasons?

These are not questions for the military alone. They require interagency and bilateral coordination to answer and convert into a strategic plan. However, the reality is that the military is engaged and often makes decisions about who is worth training, who cannot be trusted, who gets hired, which areas to cordon and search, or where a project is initiated. The military also regularly gains information on tribal power dynamics and crafts its own ad hoc models to make sense of it. Lastly and perhaps most significantly, the military is sustaining discussion on the potentiality and actuality of coopting tribal leaders. Operational and tactical commanders and their troops must deal with the dynamics of tribalism despite the lack of an integrated strategic plan.

Critiquing extremist doctrine

What of the center of gravity in the clientele version of jihadism? From what has been discussed, it appears that it is neither the person nor the legend of Osama bin Laden. If it were, one would expect to find doctrinal cohesion among the mujahideen in the camps supposedly run or supported by Al-Qaeda and between it and those remote groups who are said to respond to bin Laden's direction. Yet one finds evidence of doctrinal discord and of bin Laden's indifference to it – of his willingness to make use of even those he considers beyond the pale of Islam.[5] If it is not leadership, then perhaps Al-Qaeda's center of gravity is its aggregated capacity to project terror. However, this recourse leads to consideration of critical capabilities or resources, not center of gravity per se. Besides, Al-Qaeda's resources are of very low density and of various technological levels and are therefore relatively easy to move, conceal, replace, reschedule, or retool. There is perhaps a more subliminal dynamic at work – the possibility that the center of gravity of bin Laden's network equates to the word *qâcida* (corrupted into *qaeda*). The Arabic word has numerous meanings – basic and extended, concrete and metaphoric. It can designate base in the concrete sense of foundational or operational base; it can also designate fundamental principle. Thus, it connotes the same two dimensions, physical and moral, that were pertinent in the discussion of tribal authority. As a two-dimensional force, Al-Qaeda's critical capabilities are to uphold radical interpretations of the jihad tenet, inspire complementary actions (strikes), and covertly gain new adherents to Islamist radicalism.

Compared to the tribal case, however, the physical dimension of Al-Qaeda is diffuse – even more so than it had been – lacking geographic, institutional, or temporal consistency. Prior to the onset of Operation Enduring Freedom, the Al-Qaeda network was present in many countries in the form of mosques creating jihad-adepts and training camps generating jihadist operatives. The command center and main concentration of manpower were in Afghanistan. Consequent to Enduring Freedom and regional cooperation in the war on terror, mosque preaching was censured, and camps were aban-

---

5 Various Internet sources allude to doctrinal disagreements among the jihadists. See, for example, the section on Isam al-Turabi's testimonies in "Bin Laden's Life in Sudan—Part One," excerpt from FBIS translation of Arabic text of al-Quds al-Arabi (November 24, 2001), www. fas.org/irp/world/para/ladin-sudan.htm, 13.

doned. The militant leaders and their followers went into hiding and changed sites as needed to avoid detection. Nonetheless, capabilities in tradecraft, communications, financing, and arms procurement were conserved through better concealment techniques or modified procedures. Terrorist strikes have continued, and often it is such atrocities that first indicate presence in an area. So long as any cell can make gain for the whole movement, the effort to stop jihadist terrorism demands a long-term, wide-ranging commitment. Here again, formulation of strategy belongs in the interagency arena. The military has already shown that it has suitable assets and techniques to contribute to the cause and will likely remain engaged as long as America's will endures. However, targets such as leadership, weapons caches, and smuggling rings are in the physical dimension of Al-Qaeda. What of the moral dimension?

Many observers in the governmental, military, and media arenas have already argued for an information warfare campaign. However, the preponderance of advice calls for an external, as opposed to internal, approach — promoting tolerance, freedom, and democracy as countervalues rather than discrediting the tenets of Islamist extremism. The former approach makes little sense when the adversary's propaganda has already distorted American values into licentiousness, irresponsibility, and hypocrisy. This rejection of Westernism is buttressed by a full complement of extremist treatises and jihad lore (salvation histories, myths, and folklore that portray hero-martyrs, epic struggles, and the sense of Providence). Both sets can be targeted. However, the treatises are more vulnerable in that they lend themselves to critique on points of doctrinal validity. The jihad lore, like American frontier lore, is too embedded in the popular culture to be easily subverted. The doctrinal vulnerability, though, cannot be directly targeted. Few Americans have the knowledge to critique the tenet of takfir (as it justifies Muslims conducting jihad against other Muslims) or Sayyid Qutb's construct of the "universal Islamic concept." Even those who do would have virtually no credibility with a Muslim audience, since they would be immediately dismissed as Westerners and infidels, regardless of their credentials. The task must be shared with Muslim intellectuals who do have the credibility to critique extremist ideology yet need the technical assistance in information warfare America can offer.

One last consideration: how does the Taliban movement fit into the above scheme? The Taliban are adversaries of the United States largely because they have been, and remain, allies of Al-Qaeda. They are not, however, agents of global terrorism. They are a regional, religious-based faction that gained and lost control of most of Afghanistan. The Taliban have unity of doctrine

(Deobandist) and a high degree of ethnic homogeneity (Pashtun). Their profile is a variation of the competing allegiance dynamic graphed in figure 1. The organization, with its hard-core leadership and henchmen, retains residual support among the Pashtun tribes of Afghanistan and Pakistan. However, it continues to lose numbers through members returning to their tribal obligations and primal allegiance. The progress of Kabul's recently initiated "Reconciliation Program" should offer many examples of how wayward kinsmen are coaxed back into the tribal fold.

Afghanistan seems to offer some prospects for Iraq. However, the analogy should not be taken too far. In Afghanistan, the U.S. military had an important advantage in the initial stage of Enduring Freedom – the cooperation of a domestic ally, the Northern Alliance. This coalition not only had the necessary military and political organization to take charge of the country; it also had experience with accommodating tribalism. (During the civil war, some militias switched allegiance, according to tribal interest, just as occurred earlier in the Lebanese conflict.) The United States had no such advantage in Iraq – excluding the Kurdish autonomous zone – and thus remains challenged with developing that capacity in a new Baghdad regime. The pacification and stabilization of Iraq may consequently take longer. The bottom line is that leveraging tribalism should be critical to that effort.

There will be ample opportunity to test the above thesis because militant jihadism is likely to challenge America and its allies for some years to come. It may be an allied Muslim state, however, that ultimately leads the way against the jihadists' center of gravity. Regardless of which government leads and whether the requisite interagency approach ever becomes reality, the U.S. military must prepare to factor culture into mission planning at tactical, operational, and strategic levels. Many initiatives have been undertaken— particularly by the Army and Marine Corps, whose missions more directly engage foreign cultures. The increasing inclusion of cultural courses in service school curricula as well as cultural factors in training scenarios is a positive development. However, there are questions of proper focus and effectiveness of instructional time invested. It is important, too, to preclude easy but meaningless fixes, such as casting an exercise opposing force as Maoist-Marxist guerrillas with turbans. But where is the source of authority to rule on such issues? Perhaps the joint military community should establish a clearinghouse and staff it with specialists with genuine knowledge of indigenous customs and social dynamics, Islamic theology and social thought, and related subjects.

9

Networds: Terra Incognita and the Case for Ethnographic Intelligence

LTC Fred Renzi, USA

Author's note: What I have chosen to call "ethnographic intelligence" might be more accurately described as "ethnographic information," since much of the content involved in analyzing a hostile network will be open source. I have chosen to retain "Intelligence," however, to indicate the military utility of the content involved.

The proliferation of empowered networks makes "ethnographic intelligence" (EI) more important to the United States than ever before.[1] Among networks, Al-Qaeda is of course the most infamous, but there are several other examples from the recent past and present, such as blood-diamond and drug cartels, that lead to the conclusion that such networks will be a challenge in the foreseeable future. Given the access these networks have to expanded modern communications and transportation and, potentially, to weapons of mass destruction, they are likely to be more formidable than any adversaries we have ever faced.

Regrettably, the traditional structure of the U.S. military intelligence community and the kind of intelligence it produces aren't helping us counter this threat. As recent debate, especially in the services, attests, there is an increased demand for cultural intelligence. Retired army Major General Robert Scales has highlighted the need for what he calls cultural awareness in Iraq: "I asked a returning commander from the 3rd Infantry Division how well situational awareness (read aerial and ground intelligence technology) worked during the march to Baghdad. 'I knew where every enemy tank was dug in on the outskirts of Tallil,' he replied. 'Only problem was, my soldiers had to fight

---

1 Anna Simons and David Tucker, "Improving Human Intelligence In The War On Terrorism: The Need For An Ethnographic Capability," report submitted to Office of the Secretary of Defense for Net Assessment (2004), 5.

fanatics charging on foot or in pickups and firing AK-47s and [rocket propelled grenades]. I had perfect situational awareness. What I lacked was cultural awareness. Great technical intelligence… wrong enemy.'"[2]

I propose that we go beyond even General Scales's plea for cultural awareness and look instead at amassing EI, the type of intelligence that is key to setting policy for terra incognita. The terra in this case is the human terrain, about which too often too little is known by those who wield the instruments of national power. The United States needs EI to combat networks and conduct global counterinsurgency. This paper will therefore define EI, discuss some cases that illustrate the requirement for it, and propose a means to acquire and process it.

EI defined

According to Dr. Anna Simons of the United States Naval Postgraduate School, "What we mean by EI is information about indigenous forms of association, local means of organization, and traditional methods of mobilization. Clans, tribes, secret societies, the hawala system, religious brotherhoods, all represent indigenous or latent forms of social organization available to our adversaries throughout the non-Western, and increasingly the Western, world. These create networks that are invisible to us unless we are specifically looking for them; they come in forms with which we are not culturally familiar; and they are impossible to 'see' or monitor, let alone map, without consistent attention and the right training."[3]

Because EI is the only way to truly know a society, it is the best tool to divine the intentions of a society's members. The "Indigenous forms of association and local means of organization" are hardly alien concepts to us. Our own culture has developed what we call "social network analysis" to map these associations and forms of organization.[4] These unwritten rules and invisible (to us) connections between people form key elements of the kind of information that, according to General Scales, combat commanders are now demanding. Because these rules and connections form the "traditional methods of mobilization" used either to drum up support for or opposition to U.S. goals, they demand constant attention from the U.S. government

---

2 Robert Scales, "Culture-Centric Warfare," Proceedings (October 2004), available online at <www.military.com/newcontent/0,13190,ni_1004_culture-p1,00. Html>.

3 Simons and Tucker.

4 Valida Krebs, "An Introduction To Social Network Analysis," 2006, <www. Orgnet.com>.

and armed forces.[5] Simply put, EI constitutes the descriptions of a society that allow us to make sense of personal interactions, to trace the connections between people, to determine what is important to people, and to anticipate how they could react to certain events. With the United States no longer facing a relatively simple, monolithic enemy, our national interests are found in a confusing cauldron of different locales and societies. Each of these has its own "latent forms of social organization" that create networks we cannot see or map, and to which we may very well fall victim, unless we aggressively pursue EI.[6]

The threat: three case studies

American national interests are affected by many societies about which we may know very little. In the early 1960s, few Americans recognized the importance of the terra incognita of Vietnamese society.[7] In the 1990s, America either failed to develop, or failed to employ EI on Al-Qaeda, Afghanistan, or Iraq.[8] Today, we have little insight into which cultures or networks may soon become threats to our national interests. For this reason, America must seek to understand and develop EI on a global scale, before it is surprised by another unknown or dimly understood society or network. As a first step toward becoming more EI-smart, we might look at three illustrative cases: the blood-diamond cartel, drug trafficking syndicates, and Al-Qaeda.

The blood-diamond cartel. West Africa's blood-diamond cartel is a good example of the seemingly random mixture of networks, private armies, governments of questionable legitimacy, and social environments in conflict that plague the world today. At the core of the cartel are guerrillas in Sierra Leone who have used terror tactics to control access to diamond mines. They were assisted by the former government of Charles Taylor in Liberia, which helped launder the diamonds in Europe for money. Some of that money then went to international arms dealers who smuggled weapons to the guerrillas,

---

5 McNamara, 30-33.

6 Anonymous, *Imperial Hubris: Why The West Is Losing The War On Terror* (Washington, DC: Brassey's, 2004); Robert Baer, *See No Evil* (New York: Three Rivers Press, 2002).

7 H. Brinton Milward and Jorg Raab, "Dark Networks: The Structure, Operation, And Performance of International Drug, Terror, And Arms Trafficking Networks," paper presented at the International Conference On The Empirical Study Of Governance Management, And Performance, Barcelona, Spain, 2002, 28-39, <iigov.org/workshop/pdf/milward_and_raab. pdf>.

8 Ibid., 28.

and some went to finance international terrorists like Al-Qaeda. War, as the U.S. military has traditionally preferred to consider it-the clash of state armies and navies-has given way to a mix of crime, money, and terror executed by dark networks in league with each other and with reprehensible governments to secure profits and export terrorism. According to H. Brinton Milward and Jorg Raab, "Covert networks have come together with warlords controlling access to resources to create commodity wars. These wars are fought over control of diamonds, petroleum concessions, coca leaves, and poppies that yield narcotics, not for any real ideological or political reason."[9]

While entities like the blood-diamond cartel have heretofore not been deemed threatening to vital U.S. interests, and thus have not justified the attention of significant American assets or numbers of troops, such a presumption is overdue for reconsideration. The United States cannot afford—nor should it be inclined—to act as the world's policeman, but these unholy alliances now demand scrutiny. This is where EI enters the picture. When crime, brutality, poor governance, and terrorist financing come together, they are so enmeshed in the local social environment that only a detailed understanding of ethnographic factors can provide the basis for further identification of who and what truly threaten U.S. national interests. An understanding of the societies in which these networks roost is the indispensable bedrock upon which any further analysis rests.

Traditional military intelligence, in examining opposing formations and weapons systems, does not even speak in the same terms as those found in the blood-diamond "conflict." In Milward and Raab's words: "In the period after Taylor became president, the republic of Liberia became a nexus for many dark networks. There are linkages between various dark networks; some are more central than others are and some only loosely linked with the others."[10] Borrowed from social network analysis, terms like "network," "nexus," and "centrality" are useful concepts that allow analysts to better identify threats to American security.[11]

It is only through extensive, on-the-ground observation that latent forms of social organization and mobilization can be made apparent. When those indigenous forms of social organization are exploited by people like Charles Taylor, or become linked to external nodes such as other networks,

9   Ibid.

10   Ibid.

11   Alberto-Laslo Barabási And Eric Bonabeau, "Scale-Free Networks," *Scientific American* (May 2003): 60-69.

then EI feeds and blurs into the police style social network analysis needed to identify and counter threats to U.S. interests. In this way, EI takes the incognita out of the human terra so that the United States can craft effective, realistic policy actions.

Drug trafficking syndicates. Drug syndicates or cartels are another networked threat that will not disappear in the foreseeable future and that cannot be depicted effectively by order-of-battle-style intelligence. Phil Williams has clearly articulated the ethnic qualities that make drug trafficking a particularly opaque threat: "[M]any networks have two characteristics that make them hard to penetrate: ethnicity and language. Moreover, many of the networks use languages or dialects unfamiliar to law enforcement personnel in the host countries. Consequently, electronic surveillance efforts directed against, for example, Chinese or Nigerian drug-trafficking networks do not exist in a vacuum, but instead operate in and from ethnic communities that provide concealment and protections as well as an important source of new recruits. Some networks, such as Chinese drug-trafficking groups, are based largely on ethnicity. They are global in scope and operate according to the principle of guanxi (notions of reciprocal obligation), which can span generations and continents and provides a basis for trust and cooperation. Such networks are especially difficult for law enforcement to infiltrate. In short, drug-trafficking networks have a significant capacity to protect their information and to defend themselves against law enforcement initiatives."[12]

By themselves, drug gangs might not represent a clear and present danger to America, but they warrant study for two reasons. First, they are increasingly moving beyond mere profit-making ventures into alliances with other types of networks, such as the gun-runner and terrorist networks active in West Africa, that do pose a significant threat to the United States. Second, drug-trafficking networks provide a relevant example of how subversive groups can exploit ethnic social bonds and indigenous forms of mobilization about which we Westerners remain ignorant. Phil Williams' illustrative invocation of guanxi, which won't appear in any traditional military intelligence summary, is instructive here.

A concept of mutual obligation that can endure from generation to generation and across great distances, guanxi can be a powerful tool in the hands of a network with evil intent. Drug trafficking can be harmful enough to a society, but when it is lashed together with the trafficking of weapons,

---

12   Phil Williams, "The Nature Of Drug-Trafficking Networks," *Current History* (April 1998): 154-159.

money, and perhaps even materials of mass destruction, such racketeering does become a clear and present danger to America. A nexus of dark networks, peddling destruction in various forms, and facilitating international terrorism, becomes inordinately threatening when powered by traditional social practices such as guanxi that are invisible to states that don't do their ethnographic homework. Williams appropriately notes that these practices, or means of "indigenous mobilization," work precisely because they are embedded in an ethnic population. This is true whether the population in question inhabits an ethnic enclave in a culturally dissimilar host nation or occupies its home region. In fact, under the latter conditions, local forms of organization and means of association can become more powerful than any written law, and therefore that much more efficacious for the network using them. They can be extraordinarily effective at creating local networks. However, he who has done his ethnographic analysis stands a decent chance of neutralizing the hostile actions of a dark network or perhaps even turning the activities of the network to advantage.

Al-Qaeda. A third case that illustrates the need for EI is Al-Qaeda. In 2004, Marc Sageman wrote Understanding Terror Networks to clarify what he saw as a widespread misperception in the West about who joins these networks and why they join.

Sageman concentrates on Al-Qaeda's sub-network constituents, mapping the individual networks and partially filling in their foci, such as certain mosques.[13] Sageman obtained his information by accessing documents via friendly means, but he freely admits that his examination is limited. Sageman's main agenda is to refute the myth that terrorists such as those in Al-Qaeda are irrational psychopaths created by brainwashing impoverished Muslim youths. He contends that the majority of terrorists are educated, generally middle-class, mature adults. They are usually married, and they come from caring families with strong values. They are also believers wholly committed to the greater cause of global Salafist jihad.

According to Sageman, these people belong to four general groups in the Al-Qaeda network: the Central Staff, the Southeast Asians, the Maghreb Arabs, and the Core Arabs. The Central Staff is comprised mainly of Osama bin Laden's older compatriots, men who heard the call to jihad against the Soviet infidels in Afghanistan and who continue the fight today. The Southeast Asians are mostly disciples of two particular religious schools. The Maghreb

13 Marc Sageman, Understanding Terror Networks (Philadelphia: University of Pennsylvania Press, 2004), 143.

Arabs are first- or second-generation Arabs in France. Socially isolated, the Maghrebs have sought community ties in local mosques. The Core Arabs grew up in communal societies in Islamic lands, but became isolated and lonely as they moved away to schools or jobs.

With the exception of some Maghreb Arabs, many of Al-Qaeda's recruits have a good education and strong job skills; they have no criminal background. Sageman writes at some length about the feeling of isolation that led many of the expatriate Al-Qaeda members to seek out cliques of their own kind, and about the gradual strengthening of their religious beliefs prior to joining the jihad as a source of identity and community. He emphasizes that people join in small cliques, and that the motivation is primarily fellow-ship, and only later, worship. The cliques are not recruited as much as they seek out membership in Al-Qaeda. In the search for fellowship, some men happened upon one of the relatively few radical mosques or became embed-ded in a clique that happened to have an acquaintance in the jihadist network. Sageman debunks the theory that Al-Qaeda has recruiters in every mosque, yet he does point out the existence of a few people who know how to contact the larger group and will provide directions, travel money, and introductions to clandestine training camps. In sum, Sageman argues convincingly that our stereotypes of Al-Qaeda are dangerously misleading.

Sageman's analysis of the Al-Qaeda network has been widely quoted, yet he himself underscores the lack of available first-hand information and makes it plain that he used open-source documents, with some limited per-sonal exposure; in other words, he wrote the book without much access to EI.[14] Let us imagine what Sageman's sharp intellect would have found if he had had access to a full, well-organized range of EI from each of the four subgroups' regions. What might a dedicated core of EI specialists have discov-ered about the recruitment pattern? As an illustration, Sageman uncovered a key ethnographic point in the bond between student and teacher in Southeast Asia.[15] The active exploration of this key example of "indigenous forms of association" might have led to the two radical Southeast Asian schools much sooner. Perhaps armed with such knowledge, the governments in question could have taken more steps against the network years ago.

---

14 Ibid., vii-ix.

15 Ibid., 113-114.

Acquiring and Processing EI

To acquire ethnographic knowledge, there is no substitute for being on the scene. For the U.S. military, the structural solution to EI could be relatively easy. Some form of U.S. Military Group, or the military annex to the embassy, could become the vehicle to collect EI. While the defense attaché system is charged with overtly collecting military information and assessing the military situation in particular countries, there currently is no comprehensive effort to collect and process EI. The security assistance officers attached to U.S. country teams often obtain a fine appreciation of the cultural aspects of their host nation, but they are not charged with the responsibility to collect EI and may not always have a smooth relationship with the defense attaché (if one is even assigned).[16]

There is a relatively low-cost way to set up a system to collect EI. The United States could develop a corps of personnel dedicated to the task and base them out of a more robust military annex to our embassies. There are two key points to developing such a corps: it must be devoted exclusively to the task without distraction, and its personnel must be allowed to spend extended time in country and then be rewarded for doing so.[17] Their work could be considered a form of strategic reconnaissance, and in reconnaissance matters there is simply no substitute for being physically present on the ground. Since the ethnographic ground in question is actually a population and not necessarily terrain, a constant and near-total immersion in the local population would be the means to turn McNamara's terra incognita into a known set of "Indigenous forms of association, local means of organization, and traditional methods of mobilization."

While the most streamlined EI organization would probably combine the functions of the defense attaché and security assistance officer, such a move is not absolutely necessary.[18] The most important structural aspect is that the EI developed in country should be analyzed at the embassy, forwarded to the staff of the geographic combatant commander, and shared laterally with other relevant embassies. This kind of information sharing would make for better contingency plans, and it would create a hybrid network to counter the dark networks that profit from blood diamonds, drugs, and terror.

---

16   Kurt M. Marisa, "Consolidated Military Attaché And Security Assistance Activities: A Case For Unity of Command," *Fao Journal*, 7, 2 (December 2003): 6-11.

17   Simons and Tucker.

18   Marisa, 6-24.

A small number of Americans, usually military foreign area officers (FAOs), are already in tune with this type of work, and some have achieved a high level of excellence. There are not many of them, though, and they are not organized into a truly comprehensive system focused on the ethnographic aspects of networks. A sterling example of the capacity that the United States could build can be found in an officer named "David." On a mission with a platoon of army rangers in western Iraq to find out how foreign fighters were infiltrating the country, David traveled in Mufti. At one village, he "met a woman with facial tattoos that marked her as her husband's property. As they chatted, the pale-skinned, sandy-haired North Carolina native imitated her dry, throaty way of speaking. 'You are Bedu, too,' she exclaimed with delight." From her and the other Bedouins, David finds out that the foreign fighters are using local smuggling routes "to move people, guns, and money. Many of the paths were marked with small piles of bleached rocks that were identical to those David had seen a year earlier while serving in Yemen."[19]

David gained access and operational information by using ethnographic knowledge. The deeper that personnel like David dig into local society, the better their ability to assess which groups threaten the United States and which should be left alone. If America could build a healthy corps of people like David, based out of each U.S. embassy in the world, then our nation could identify those networks that, in Simons's formulation, are "invisible to us unless we are specifically looking for them; [and that] come in forms with which we are not culturally familiar."

Sadly, there aren't nearly enough Davids in the military. The army has about 1,000 FAOs, but most of them are in Europe. A mere 145 are focused on the Middle east, and even that number can be deceptive because a FAO's duties include many things that aren't related to EI, such as protocol for visits and administrative duties.[20] Certainly, one solution to the growing threats from networks would be to produce more Davids and reward them for extensive time on the ground exclusively focused on the development of EI.

The benefits to be derived from such a corps would be tremendous. Consider, for example, the impact good EI could have had on the war plan for Iraq. There has been much discussion of late about how American forces did not really understand the Iraq's tribal networks, a failure that contributed to the difficulties we are currently facing. With the "consistent attention and the

---

19 Greg Jaffe, "In Iraq, One Officer Uses Cultural Skill To Fight Insurgents," *Wall Street Journal*, 15 November 2005, 15.

20 Ibid.

right training" Simons has prescribed, knowledge like this could have been built into contingency plans and then updated in the regular two-year plan review cycle to insure currency. Ethnographic understanding could have allowed U.S. forces in Iraq to use tribal networks to advantage from the outset; they would not have had to figure things out for themselves, as Lieutenant Colonel Tim Ryan did: "The key is a truce brokered by the national league of Sheiks and tribal leaders and U.S. Army LTC Tim Ryan, the 1st Cavalry Division officer responsible for Abu Ghraib—a Sunni triangle town west of Baghdad and a hotbed of the insurgency. Under the agreement, Ryan now meets regularly with tribal leaders and provides them with lists of residents suspected of taking part in attacks. The sheiks and their subordinate local clan leaders then promise to keep their kinsmen in line. 'They [the sheiks] do have a lot of influence. To ignore that is to ignore 6,000 years of the way business has been done here.'"[21]

EI that might lead to beneficial relations with local power figures, along the lines of the one between Ryan and the sheiks, could be developed from each U.S. embassy around the clock in peacetime to inform contingency plans and enable activity against the dark networks that seek to harm America. In some places, such as pre-war Iraq or in outright killing fields similar to a blood-diamond zone, Washington will judge the presence of an embassy to be too dangerous, but in the absence of an on-site embassy, personnel can be invested in the surrounding embassies to glean as much EI as possible through borders that are often porous.

The Broken Windows theory of criminologists James Q. Wilson and George Kelling suggests that we might reap another benefit from establishing an American ethnographic counter-network in surrounding, linked embassies.[22] The essence of the theory is that if a building has a broken window that remains unfixed, then people will assume that no one is in charge or cares; as a result, they will do whatever they wish to the place – the broken window will invite vandalism, graffiti, and so on. Once these acts of disorder commence, crime becomes contagious, like a fashion trend or virus. A more robust military annex to an embassy and a low-key, constant interest in overt ethnographic matters would show that the United States cares and is indeed watching. Perhaps this constant attention would serve to subtly constrict the

21 Ashraf Khalil, "Teaming up with tribes to try to quell insurgents," Los Angeles Times, 21 June 2004, A8.

22 Malcolm Gladwell, The Tipping Point: How Little Things can Make a Big Difference (New York: Little, Brown, and Co., 2000), 140-146.

amount of safe haven space available for dark networks. The overt information gathered by military ethnographers could complement the covert work done by the CIA (and vice versa).

U.S. citizens, at least intuitively, have always recognized the presence of networks in society, from family ties to economic relationships, indeed, to the very structure of daily life. The law enforcement community has long since recognized and acted against domestic criminal and extremist variants of these networks. However, the U.S. government and military have had a difficult time coming to grips with networks like Al-Qaeda. It took the shock of the September 11th attacks to galvanize national attention on terrorist networks, and the ensuing years of struggle to grasp that terror networks can be more than ideologically motivated, and that they can flourish in the nexus of crime, drugs, weapons trafficking, money laundering, and a host of other lethal activities.

Terrorism can take many guises, and it blends very well into the cauldron of dark phenomena like blood diamonds, drug trafficking networks, and Al-Qaeda. The United States desperately needs a counter-network to fight the dark networks now surfacing across the globe. Ethnographic intelligence can empower the daily fight against dark networks, and it can help formulate contingency plans that are based on a truly accurate portrayal of the most essential terrain – the human mind. United States policymakers must not commit us ever again to terra incognita. The nation must invest in specialized people who can pay "constant attention" to "indigenous forms of association and mobilization," so that we can see and map the human terrain.

10

The Human Terrain System: A CORDS for the 21st Century

Jacob Kipp, Lester Grau, Karl Prinslow, and CPT Don Smith, USA

In accurately defining the contextual and cultural population of the task force battlespace, it became rapidly apparent that we needed to develop a keen understanding of demographics as well as the cultural intricacies that drive the Iraqi population.
—Major General Peter W. Chiarelli, Commander, 1st Cavalry Division, Baghdad, 2004-2005[1]

Conducting military operations in a low-intensity conflict without ethnographic and cultural intelligence is like building a house without using your thumbs: it is a wasteful, clumsy, and unnecessarily slow process at best, with a high probability for frustration and failure. But while waste on a building site means merely loss of time and materials, waste on the battlefield means loss of life, both civilian and military, with high potential for failure having grave geopolitical consequences to the loser.

Despite these potential negative consequences, the U.S. military has not always made the necessary effort to understand the foreign cultures and societies in which it intended to conduct military operations. As a result, it has not always done a good job of dealing with the cultural environment within which it eventually found itself. Similarly, its units have not always done a good job in transmitting necessary local cultural information to follow-on forces attempting to conduct phase iv operations (those operations aimed at stabilizing an area of operations in the aftermath of major combat).

Many of the principal challenges we face in Operations Iraqi Freedom and Enduring Freedom (OIF and OEF) stem from just such initial institutional disregard for the necessity to understand the people among whom our forces

---

1 MG Peter W. Chiarelli and MAJ Patrick R. Michaelis, "Winning the Peace: The Requirement for Full-Spectrum Operations," *Military Review* (July-August 2005), 5.

operate as well as the cultural characteristics and propensities of the enemies we now fight. To help address these shortcomings in cultural knowledge and capabilities, the Foreign Military Studies Office (FMSO), a U.S. Army Training and Doctrine command (TRADOC) organization that supports the Combined Arms Center at Fort Leavenworth, Kansas, is overseeing the creation of the human terrain system (HTS). This system is being specifically designed to address cultural awareness shortcomings at the operational and tactical levels by giving brigade commanders an organic capability to help understand and deal with "human terrain" – the social, ethnographic, cultural, economic, and political elements of the people among whom a force is operating.[2] So that U.S. forces can operate more effectively in the human terrain in which insurgents live and function, HTS will provide deployed brigade commanders and their staffs direct social-science support in the form of ethnographic and social research, cultural information research, and social data analysis that can be employed as part of the military decision-making process.

The core building block of the system will be a five-person Human Terrain Team (HTT) that will be embedded in each forward-deployed brigade or regimental staff. The HTT will provide the commander with experienced officers, NCOs, and civilian social scientists trained and skilled in cultural data research and analysis. The specific roles and functions of HTT members and supporting organizations are discussed below.

To augment the brigade commander's direct support, HTS will have reachback connectivity to a network of subject-matter experts now being assembled from throughout the department of defense, the interagency domain, and academia. This network will be managed by a centralized information-clearinghouse unit nested in FMSO. At the same time, to overcome the kinds of problems now typically encountered when in-place units attempt to transfer knowledge about their area of operations upon relief in place, HTS

---

2 The concept for the current Human Terrain System was suggested by Montgomery McFate Ph.D., J.D., and Andrea Jackson as described in their article, "An Organizational Solution for DoD's Cultural Knowledge Needs," Military Review (July-august 2005), 1821. Most of the practical work to implement the concept under the title Human Terrain System was done by Cpt Don Smith, U.S. Army reserve, of the Foreign Military Studies Office, between July 2005 and August 2006. Under this concept, "human terrain" can be defined as the human population and society in the operational environment (area of operations) as defined and characterized by sociocultural, anthropologic, and ethnographic data and other non-geophysical information about that human population and society. Human terrain information is open-source derived, unclassified, referenced (geospatially, relationally, and temporally) information. It includes the situational roles, goals, relationships, and rules of behavior of an operationally relevant group or individual.

will provide for the complete transfer of HTT personnel together with the HTT database to the incoming commander upon transfer of authority. This will give the incoming commander and unit immediate "institutional memory" about the people and culture of its area of operations.

Five HTTs will deploy from Fort Leavenworth to Afghanistan and Iraq beginning in the fall of 2006 to provide proof-of-concept for the HTS. If they are successful, an HTT will eventually be assigned to each deployed brigade or regimental combat team.

Why we need HTS – history

Cultural awareness will not necessarily always enable us to predict what the enemy and noncombatants will do, but it will help us better understand what motivates them, what is important to the host nation in which we serve, and how we can either elicit the support of the population or at least diminish their support and aid to the enemy.
–Major General Benjamin C. Freakley, Commanding General, CJTF-76, Afghanistan, 2006[3]

The many complex and unexpected issues resulting from lack of cultural knowledge have often been extraordinarily challenging for newly deployed commanders and their soldiers, especially in insurgent environments like those of OIF and OEF. To address recent challenges, many military thinkers have independently sought answers by studying practices and procedures from previous historical experiences. Consequently, the writings of T.E. Lawrence and David Galula have become standard reading for those searching for answers to the current insurgencies.[4] Interest has also been rekindled in the U.S. Marine Corps's Small Wars Manual, a volume first published in 1940 that outlines doctrine the Corps developed for counterinsurgency in other eras.[5] Other thinkers have reexamined the basics of more recent counterinsurgency practices, in Vietnam and elsewhere, in the search for appropriate and currently applicable counterinsurgency measures.[6] Still others have gone back to

---

3 MG Benjamin C. Freakley, Infantry 94, 2 (March-April 2005), 2.

4 See T. E. Lawrence, *Seven Pillars of Wisdom* (New York: Anchor Books, 1991); and David Galula, *Counterinsurgency Warfare: Theory and Practice* (New York: Praeger Press, 1964).

5 United States Marine Corps Small Wars Manual, 1940 (Manhattan, KS: Sunflower University Press), 1-2.

6 Lester Grau and Geoffrey Demarest, "Maginot Line or Fort Apache? Using Forts to Shape the Counterinsurgency Battlefield," *Military Review* (November-December 2005): 35-40.

the lessons of British imperial and French colonial experience.[7]

What has emerged overall from these varied examinations of the historical record of insurgency is a broad consensus that civil society in Iraq and Afghanistan – as in past insurgencies – constitutes the real center of gravity. The current insurgencies in the middle east are manifestations of the unmet expectations and desires of large segments of the Iraqi and Afghani populations. Disappointed by their unrequited aspirations, the people tolerate and even support the presence of insurgents, thereby making insurgency possible. Such conclusions logically demand that past experience guide our understanding of how best to meet, in a manner that supports our own military objectives, the expectations and desires of the people at the heart of such struggles. And, to truly understand such expectations and desires, it is imperative to view them from the perspective of the cultures in which the insurgencies are being waged.

Learning from Vietnam

History has shown that insurgency is a complex form of armed struggle that can only be dealt with effectively if the counterinsurgent makes an effort to understand the conflict from its origin, through its evolutionary stages of development, down to its current situation. Most insurgent wars have been inherently political in nature, and therefore share the characteristic of having been decided by one side or the other's ability to finally win the allegiance of the general civil population in the conflict area.

In contrast, however tempting it may be to advocate "draining the swamp" by force as a solution to insurgency (i.e., denying the insurgency support by uprooting or terrorizing the local population), such policies have historically only increased popular resentment, eroded popular trust, and stimulated the indigenous recruitment of additional insurgents.

While history offers many examples of insurgencies worthy of study, the HTS concept has been largely inspired by lessons drawn from the U.S. experience in Vietnam. During the Vietnam conflict, U.S. Armed Forces essentially fought two different wars: one a conventional war against regular

---

7 See, for example, Andrew M. Roe, "To Create a Stable Afghanistan: Provisional Reconstruction Teams, Good Governance and a Splash of History," *Military Review* (November-December 2005), 20-26; LTC James D. Campbell, "French Algeria and British Northern Ireland: Legitimacy and the Rule of Law in Low-Intensity Conflict," *Military Review* (March-April 2005), 2-5; and Col. (retired) Henri Bore, "Cultural Awareness and Irregular Warfare: French Army Experience in Africa," *Military Review* (July-August 2006), 108-111.

North Vietnamese formations; the other an insurgency war against guerrillas who, for a long time, moved freely throughout the area of operations because they enjoyed the support of a significant number of the rural South Vietnamese people. The record reveals that U.S. counterinsurgency efforts in the early part of the conflict were severely hobbled by a lack of understanding of, or appreciation for, Vietnamese culture, and a paucity of cultural skills, especially language ability.

Subsequently, among the many weapons brought to bear against the insurgency in South Vietnam during the course of the war, perhaps the most effective was one that involved South Vietnamese forces backed by advisors from the Civil Operations and Revolutionary Development Support (CORDS) program, a project administered jointly by the South Vietnamese Government and the Military Assistance command, Vietnam (MACV). Implemented under the Johnson administration, the CORDS program specifically matched focused intelligence collection with direct action and integrated synchronized activities aimed at winning the "hearts and minds" of the South Vietnamese. CORDS was premised on a belief that the war would be ultimately won or lost not on the battlefield, but in the struggle for the loyalty of the people.[8]

With CORDS, intelligence collection and civil military operations were consolidated under a single civilian head, in order to shift the focus of military operations from defeating the North Vietnamese army and regional communist guerrillas by direct military force, to working with the South Vietnamese to gather human and cultural intelligence and to develop economic and social programs. These latter programs aimed to undermine indigenous support for the communist forces.

William Colby, one of the architects of this strategy, later blamed the final loss in Vietnam on failure to fully implement the CORDS strategy. Colby asserted that the "major error of the Americans in Vietnam was insisting upon fighting an American style military war against an enemy who, through the early years of the war, was fighting his style of people's war at the level of the population."[9] Colby asserted that efforts to transform rural life through

---

8   See, for example, Dale Andrade, *Ashes to Ashes: The Phoenix Program and the Vietnam War* (Lexington, Ma: lexington Books, 1990); Ralph W. Johnson, "Phoenix/Phung Hoang: a Study of Wartime Intelligence Management" (Ph.D. Diss., the American University, 1985); Dale Andrade and James H. Willbanks, "Cords/Phoenix: Counterinsurgency Lessons from Vietnam for the Future," *Military Review* (March-April 2006): 9-23; and MAJ Ross Coffey, "Revisiting CORDS: the Need for Unity of Effort to Secure Victory in Iraq," *Military Review* (March-April 2006): 35-41.

9   William Colby with James McCargar, *Lost Victory: A First-Hand Account of America's Sixteen Year*

economic development would create the conditions necessary to foster peace and stability. Such development, he maintained, would counter any appeal the terrorists might have for the people by creating local opportunities for the people to exercise real freedoms within their own institutions and values.[10]

More recent work appears to validate Colby's assessment. Robert K. Brigham stresses this point in a study assessing the South Vietnamese army and its linkages to its own society – the society from which the army had to draw its resources and its legitimacy.[11] Colby's views are further supported by the work of James H. Willbanks. In his recent treatment of Vietnamization, Willbanks addresses the tension between defeating the opposing regular force and pacifying the south in the final stages of that war (1968-1975). He underscores the linkage between pacification and Vietnamization, and argues that the former contributed to the overall stability of rural South Vietnam.[12]

Despite CORDS' shortcomings (the overall success of the program is still heatedly debated by historians), it is hard to argue with the statistics from that era. Where CORDS was effectively implemented, enemy activity declined sharply. In memoirs and records opened in the aftermath of the conflict, North Vietnamese leaders repeatedly express their concern about the effectiveness of the CORDS program in impeding both their operational and subversion campaigns.[13]

A key feature leading to the success of CORDS was an effective information collection and reporting system that focused on factors essential for the promotion of security, economic development, governance, and the provision of needed government services down to the hamlet level. Cultural, economic, and ethnographic reports were paralleled by monthly reports on the training, equipment, morale, and readiness of Vietnamese Armed Forces from the separate platoon level to the highest echelons.[14] Though imperfect, the systematic collection of such information gave both the South Vietnamese Government and MACV sufficient situational awareness, at the granular level

---

War in Vietnam (Chicago & New York: Contemporary Books, 1989), 175-192.

10 Amartya Sen, *Development as Freedom* (New York: Anchor Books, 2000), 227-248.

11 Robert K. Brigham, *ARVN: Life and Death of the South Vietnamese Army* (Lawrence: University Press of Kansas, 2006), 1-26.

12 James H. Willbanks, *Abandoning Vietnam: How America Left and South Vietnam Lost Its War* (Lawrence: University Press of Kansas, 2004), 56-58.

13 Andrade and Willbanks, 21-22.

14 Ibid., 14-17.

of detail needed, to cope effectively with many areas dominated by insurgents. The major problem with CORDS appears to have been that it was started too late and ended too soon.

Regardless, the Vietnam-era CORDS experience provides many important lessons to guide the development of an effective cultural intelligence program, one that can support tactical-and operational-level commanders today.

Among the most significant deficiencies evident in the otherwise effective CORDS program was that it had limited reachback capability. This meant that cords operators had to rely mainly upon the program's own independently developed databases and sources for information. CORDS was not structured or resourced to take full advantage of the massive U.S. capabilities for cultural and social research and analysis that would have enabled even greater effectiveness in dealing with the culturally diverse environment of Vietnam. Instead, CORDS advisory teams were left largely to their own devices to invent collection systems and methods for storing and analyzing their own data. HTS will not suffer such shortfalls in capability.

Why we need HTS today

In the current climate, there is broad agreement among operators and researchers that many, if not most, of the challenges we face in Iraq and Afghanistan have resulted from our failure early on to understand the cultures in which coalition forces were working. In other words, failing to heed the lessons of Vietnam and CORDS, we did not take the steps necessary to deal appropriately with the insurgencies within the context of their unique cultural environments. Moreover, there appears to be general agreement that whatever notable successes we have had in specific localities closely correlate with proactive efforts by coalition units to understand and respect the culture. By conducting operations that took indigenous cultural norms into account, those units garnered support for coalition objectives.

Yet, current intelligence systems and organizations still remain primarily structured to support commanders in physical combat. They are engineered to collect traditional elements of information like order of battle, enemy dispositions and estimated capabilities, and friendly and neutral capabilities for actual combat. Generally, such data is maintained in automated databases and arrayed on computer screens that depict enemy forces, friendly forces, communications nodes, key logistics facilities, and the like.

But, as the current conflicts have moved further away from combat involving regular formations and heavy maneuver warfare, and more toward insurgency operations with fragile stability operations requirements, it is now apparent that the technical information required for high-intensity conflict has diminished in importance relative to the requirement for the kind of ethnographic, economic, and cultural information needed to stabilize a polity and transfer power to an indigenous government. Irrespective, today, commanders arriving in their areas of operation are routinely left to fend for themselves in inventing their own systems and methodologies for researching and analyzing such data. Developing a system and processes requires the expenditure of enormous amounts of precious time and involves a great deal of trial and error, together with a steep learning curve. The resulting database is generally accomplished through ad hoc rearrangement of the staff. Nor are these homegrown databases formally linked to other databases to allow the seamless sharing of information or the archiving of data for broader use within the Army. Moreover, the database and institutional memory that go with it are not effectively transferred to relieving units upon redeployment. As a result, new commanders entering the area of operations usually must start again from scratch, developing their own system for researching and analyzing cultural data.

Consequently, it is glaringly apparent that commanders need a culturally oriented counterpart to tactical intelligence systems to provide them with a similarly detailed, similarly comprehensive cultural picture of their areas of operations.

HTS aims to mitigate these problems by providing commanders with a comprehensive cultural information research system that will be the analogue to traditional military intelligence systems. It will fill the cultural knowledge void by gathering ethnographic, economic, and cultural data pertaining to the battlefield and by providing the means to array it in various configurations to support analysis and decisionmaking. Moreover, the forward deployed brigade-level elements upon which the system is based will have reachback capability for research. Additionally, the whole database and institutional memory will be transferred in total to successive commanders upon unit rotation, providing for needed continuity of situational awareness.

A closer look at HTS

In its current conception, HTS is built upon seven components, or "pillars": human terrain teams (HTTs), reachback research cells, subject-matter expert networks, a tool kit, techniques, human terrain information, and specialized training.

Each HTT will be comprised of experienced cultural advisors familiar with the area in which the commander will be operating. The actual experts on the ground, these advisors will be in direct support of a brigade commander. All will have experience in organizing and conducting ethnographic research in a specific area of responsibility, and they will work in conjunction with other social-science researchers. HTTs will be embedded in brigade combat teams, providing commanders with an organic capability to gather, process, and interpret relevant cultural data. In addition to maintaining the brigade's cultural databases by gathering and updating data, HTTs will also conduct specific information research and analysis as tasked by the brigade commander.

Teams will consist of five members: a leader, a cultural analyst, a regional studies analyst, a human terrain research manager, and a human terrain analyst.

• The HTT leader will be the commander's principal human terrain advisor, responsible for supervising the team's efforts and helping integrate data into the staff decision process. He or she will be a major or lieutenant colonel and a staff college graduate, and will have spent time as a principal brigade staff officer.

• The cultural analyst will advise the HTT and brigade staff and conduct or manage ethnographic and social-science research and analysis in the brigade's area of operations. The analyst will be a qualified cultural anthropologist or sociologist competent with geographical imaging software and fluent enough in the local language to perform field research. Priority selection will go to those who have published, studied, lived, and taught in the region.

• The regional studies analyst will have qualifications and skills similar to the cultural analyst.

• The human terrain research manager will have a military background in tactical intelligence. The manager will integrate the human terrain research plan with the unit intelligence collection effort, will debrief patrols, and will interact with other agencies and organizations.

• The human terrain analyst will also have a military intelligence background and be a trained debriefer. He or she will be the primary human terrain data researcher, will debrief patrols, and will interact with other agencies and organizations. The HTT will be responsible to the brigade commander for three deliverables:

•   A constantly updated, user-friendly ethnographic and sociocultural database of the area of operations that can provide the commander data maps showing specific ethnographic or cultural features. The HTT's tool kit is mapping Human terrain (map-Ht) software, an automated database and presentation tool that allows teams to gather, store, manipulate, and provide cultural data from hundreds of categories. Data will cover such subjects as key regional personalities, social structures, links between clans and families, economic issues, public communications, agricultural production, and the like. The data compiled and archived will be transferred to follow-on units. Moreover, although map-Ht will be operated by the HTTs, the system will regularly transfer data to rear elements for storage in a larger archive, to allow for more advanced analysis and wider use by the military and other government agencies.

•   The ability to direct focused study on cultural or ethnographic issues of specific concern to the commander.

•   A reachback link to a central research facility in the United States that draws on government and academic sources to answer any cultural or ethnographic questions the commander or his staff might have.

Finally, as previously noted, the team and database will not displace when a commander or unit departs upon change of responsibility. Instead, the HTT will transfer in its entirety to the incoming commander and unit.

Reachback specifics

To provide the reachback that CORDS lacked, an organization called the HTS Reachback Research Center (RRC) will be established as part of the Foreign Military Studies Office at Fort Leavenworth. All HTTs will have direct connectivity with the RRC.

Initially, the RRC will have 14 researchers, all experts in the cultural and ethnographic characteristics of the geographic area they support. The RRC will systematically receive information from deployed HTTs through the map-Ht system. Data will be collated, catalogued, and placed into a central database. The RRC will also be able to conduct additional analysis in support of forward HTTs.

The RRC's main purpose is to help HTTs answer forward-deployed commanders' specific requests for information. Apart from its own institutional expertise, the RRC will be able to access a network of researchers throughout the government and academia to conduct research and get answers. RRC researchers will also constitute the primary pool from which replacements for forward HTTs will be drawn. RRC personnel will periodically rotate into theater to serve tours as forward HTT members. They will be designated to reinforce in-theater HTTs during an emergency or in a surge period, as required by a brigade commander.

Overall system

In addition to the capabilities the HTS offers to brigade commanders and other decisionmakers in given areas of operation, the data it compiles will be available for the training, modeling, and simulation communities to better support deploying forces in their mission rehearsal exercise scenario development. Other U.S. Government agencies will also have access to the central database. And finally, to facilitate economic development and security, the compiled databases will eventually be turned over to the new governments of Iraq and Afghanistan to enable them to more fully exercise sovereignty over their territory and to assist with economic development.

Getting the data

Most civilian and military education is based on unclassified or open-source information derived from the social sciences. Similarly, most cultural information about populations is unclassified. To ensure that any data obtained through the HTS does not become unnecessarily fettered or made inaccessible to the large numbers of soldiers and civilians routinely involved in stability operations, the information and databases assembled by the HTS will be unclassified.

Many grounds for optimism

To date, although our brigades have performed with heroism and distinction in Iraq and Afghanistan, lack of cultural knowledge and language capabilities appear to have been major common factors standing in the way of optimal success. With the introduction of the HTS and its human terrain teams, future deploying brigades will get a running start once they enter theater. They will be culturally empowered, able to key on the people and so prosecute counternotes insurgency as Lawrence, Galula, and other practitioners have prescribed – not by fire and maneuver, but by winning hearts and minds. In turn, the army, our nation, and the people of Iraq and Afghanistan will benefit from the fielding of this powerful new instrument for conducting stability operations and reconstruction.

III
One Size Does Not Fit All

11

Avoiding the Cookie Cutter Approach to Culture: Lessons Learned from Operations in East Africa

MAJ Christopher H. Varhola, USAR and
LTC Laura R. Varhola, USA

several years ago, a group of cease-fire monitors preparing to go to the Nuba Mountains in Sudan received a situation briefing in the Pentagon. At the conclusion of the briefing, one monitor asked about crime and economic violence in the area. The briefing officer, a colonel in the Army, patiently explained that the conflict in the Sudan was between Muslims and Christians and that crime was not a concern. His response, which reflected a common approach to examining conflict, underscored the need to integrate cultural understanding into the spectrum of military operations. The reality in the Sudan and elsewhere is that political, economic, and religious factors cannot be examined in isolation. In that area of the Sudan, for instance, competition between herders and farmers had political, religious, and military dimensions. The economic tension framed much of the conflict, and escalating economic violence was the single largest threat to the cease-fire.

Culture has been described as "multiple discourses, occasionally coming together in large systemic configuration, but more often coexisting within dynamic fields of interaction and conflict."[1] Culture is so broad that we cannot isolate it and study it apart from other societal factors such as history, economics, politics, religion, and relationships ranging from local to international. But in both military history and counterinsurgency literature, references to culture and regional understanding too often consist of a single line or paragraph stating that such knowledge is critical for success. In the

---

1 Nicholas Dirks, Geoff Eley, and Sherry Ortner, *Culture/Power/History* (Princeton, NJ: Princeton University Press, 1994), 4.

past, one-hour cultural briefs conducted during preparation for deployment often misrepresented the culture and diminished its importance in planning operations. Now, largely because of challenges in Iraq, there is a growing recognition of the need for cultural awareness and understanding in the military. Lessons learned in Iraq include the need for:

- Continuity of personnel and institutional knowledge in each region.
- Cultural training in our educational institutions.
- Diversity in language capabilities.
- Socioeconomic analysis conducted during the planning process by regional specialists.
- Timely reachback to sector specialists.

In 2002 the U.S. military established the Combined Joint Task Force-Horn of Africa (CJTF-HOA) in Djibouti for the purpose of "detecting, disrupting, and ultimately defeating transnational terrorist groups operating in the region."[2] Part of its mission involves economic assistance in the form of civil-military operations to reduce the conditions of poverty that help foster terrorism. Implied in this endeavor is an understanding of complex socioeconomic and cultural factors that influence the behavior and beliefs of peoples throughout the Horn of Africa and parts of East Africa.

Inadequate preparation and planning

Despite the lessons learned in Iraq, operations like those ongoing in Kenya and Tanzania are marked by high personnel turnover. Moreover, most of the personnel deployed have received little or no training on the region, have no Swahili language ability, and do not have a chain of command insisting that they learn the indigenous language in situ. To compound the problem further, few of those who plan the operations have been to the countries involved, and, even if the planning staff includes a section of regional specialists, the section usually has little influence on other staff sections. We can attribute the latter deficiency to the way military staffs typically work; that is, they tend to operate independently and focus on a functional area rather than integrating all aspects of local and regional variations into their operational plan. Regulations, standard operating procedures, models, and guidelines developed in other contexts reinforce this tendency. As a result, the staff develops the plan in a vacuum with little regard for the importance of regional concerns and specificities.

---

2 Mission statement, Combined Joint Task Force-Horn of Africa, www.hoa.centcom.mil.

Mistaking the power of tribal identity. It is very common in Iraq to hear American military personnel state that Iraqi society is tribal, and that if one understands tribes, then one understands Iraq. The same thinking is common in East Africa. Because war often involves the complete breakdown of political and economic structures, theories about the resurgence of primal religious and ethno-tribal identities rise to prominence. These theories focus on cross-cultural interactions and insist that some basic interactions supplant other forms of interaction. This analysis is tempting in its simplicity, but it is wrongheaded. The variable role of tribal identity is certainly important within the shifting mix of other factors such as race, religion, nationality, history, mode of livelihood, and locality; however, none of these factors can be examined in isolation from the other factors or under conditions that stress one factor over others.

Tribal identities may play a less obvious role in peacetime engagement activities because these operations usually occur in sovereign countries with functioning governments and judicial systems that might hold greater sway than cultural and ethnic concerns. Nevertheless, cultural factors play an important role in governmental and societal structures. Accordingly, each staff section must consider them during planning and execution. This simplistic statement may be axiomatic, but its application is complex.

Overlooking cultural complexity. The cookie-cutter approach to incorporating culture in operational planning for humanitarian and other peacetime operations is simplistic; it disregards the complex reasons why people choose terror as a form of action. For example, consider the August 2006 press conference in Tanzania at which a senior U.S. military commander declared that the U.S. military was in Tanzania "going after the conditions that foster terrorism." Tanzanians were perplexed by the commander's comments, and a reporter from the Associated Press found them amusing and went around asking Tanzanians if they had seen any terrorists recently. Tanzanians greatly appreciated the U.S military effort, but the reason given for providing assistance did not enhance critical ties of trust to the degree they could have.[3]

The politicization of discontent born from poverty and social oppression is nothing new. It has long been part of the rationale behind the U.S. Agency for International Development and its counterparts in foreign governments. Saying that poverty alone causes terrorism simplifies complex situations and ignores a bevy of other factors besides gross domestic product that

3 "U.S. Builds Clinic to Win Hearts and Minds of Tanzanians," *This Day* (Dar es Salaam, Tanzania), 19 August 2006.

affect social conditions and attitudes. The commander in Tanzania conducting the press conference wanted to publicize U.S. military humanitarian-assistance activities. But his comments, obviously linking U.S. actions to fighting terrorists, actually lessened the effectiveness of the operation: they drew attention to the fact that American forces were in Tanzania to advance U.S. national interests, not to improve the welfare of the Tanzanian people. The commander's comments revealed his staff's limited focus and lack of knowledge of the intricacies of Tanzanian rural areas.

Dubious public-affairs efforts. Military public affairs officers are supposed to be specialists in dealing with the media, but without experience in a given region, they often default to the idea that the more press there is, the better. However, if the purpose of an operation is to improve social conditions, thereby reducing an area's potential as a breeding ground for terrorists, then publicizing the action would be largely unnecessary and perhaps even counterproductive. Local news passed by word-of-mouth is sufficient to inform the target audience about the U.S. effort and to convey the idea that Americans are undertaking humanitarian assistance for more than the sake of immediate attention and gain. Unfortunately, U.S. military and State Department personnel often do only a one- or two-year tour of duty, which limits their impact and the number of projects they can effect. It is understandable that they want to publicize the actions they do undertake, but unreflective publicity can make it appear that the United States is involved in high-visibility, flash-in-the-pan actions, not long-term programs. Informing the national and international news media about these operations invites criticism because it opens U.S. actions up to a larger audience, one that might link the operations to "militaristic" or "imperialistic" U.S. actions elsewhere in the world. This is less the case when publicity is limited to the local level.

Misunderstanding religious influence. Perceptions that rural areas in Tanzania are potential breeding grounds for Islamic extremism are not necessarily wrong, but they generally ignore local religions, paths of development, civic attitudes, and the popularity and accessibility of elected government officials. In the district where the commander made his remarks, there is a historical blend of Islam and Christianity (the latter mainly Catholic and Anglican) under a larger African cultural umbrella. This syncretic religious mix recognizes the role and power of spirits and magic, as well as the influence of family ancestors, in contemporary life. It also fosters a religious tolerance that promotes coexistence and economic networking. Throughout the locality, interfaith marriages are common, as are conversions from Islam to Christianity and vice versa (with gender playing no role).

Lately, however, an influx of external, less tolerant religious influence has been challenging the status quo. Specifically, there is a growing number of Pentecostalists who have declared that placating the spirits of one's ancestors is a form of devil worship and that Muslims are barred from heaven because they do not accept Jesus as a god. But Muslims in the area have refuted the Pentecostalists' attempt to divide the community. By deeming the Pentecostalists to be heretics who worship Jesus instead of God — and not merely a different Christian sect of the same (syncretic) religion — they have effectively expelled the newcomers from the larger community. The theological specifics of the Christian Trinity have proven to be less important than maintaining a system that allows for peaceful coexistence. Similarly, extreme Muslim views that do not accommodate local beliefs and allow conversion to Christianity are unlikely to resonate with these Tanzanians. Obviously, this greatly affects the area's potential to breed terrorists. We should incorporate this fact into American civil-military strategies.

Ignoring economic and power relations. The commander's comments also ignored civic identities and modes of livelihood that affect attitudes and proclivities toward supporting or using violence. Political opposition to the United States in the area is limited, but where it does exist, it must be placed in socioeconomic context, not be taken at face value — appearances can be misleading. For example, a majority of residents in another, overwhelmingly Muslim, village in the same district declared their hatred for America and stated that no American was welcome there. These villagers couched their views in political and religious rhetoric, but in this case, politics and religion were less important than economics. The village sits on the coast, and its residents were smuggling marijuana, mangrove poles, and poached meat to Zanzibar and the Middle East. The attitudes they espoused were less political than pragmatic: they wanted to minimize outside attention to the area because it would disrupt their ongoing illegal enterprises.

Likewise, on a recent visit to Bagamoyo District, we observed a large number of cattle herders. These people had recently moved into the area because of a drought in their traditional grazing lands. Their presence is a source of tension, and conflict with farmers in the district is common. Consequently, U.S. civic action to provide veterinary services to the herders' cows might seem an obvious course of action, but it would likely anger the indigenous residents of the area and generate ill will toward the United States.

One fallacy shared by Americans and many Westerners is the belief that civic action projects are always positive and relatively simple to execute. The idea that local populations must perceive such activities as beneficial is just not true. In the former colonial countries of East Africa, religion was a tool for colonization, and the motto "Uhuru and Kujitegemea" (Freedom and Self-Reliance) indicates East Africa's resolve to avoid a repeat of the dependency relationships of unequal exchange that characterized the colonial era. Even if development is correctly billed as an effort to win hearts and minds, it is not always seen as a benign force. The United States cannot gain the acceptance of a population simply by spending money on social projects. On the contrary, the population often regards such expenditures as another way for developed nations to advance their national agendas and diminish African sovereignty.

Developmental assistance is also frequently portrayed as a cover for military and intelligence operations. For instance, several months ago, Tanzanian and Kenyan newspaper articles discussed a U.S. military "top secret plan" to fight terrorism. The articles stated that Army coordination elements and military liaison elements, composed of highly trained Green Berets proficient in local languages, were operating under the cover of humanitarian projects to collect intelligence and infiltrate terror networks.[4] One can see how easy it is to associate contemporary civil-military operations with covert military operations. The U.S. military must establish priorities and guidelines with regard to conducting these operations.

Who should do culture?

Understanding the role culture plays in society is neither an easy task nor one for which military. One fallacy shared by Americans and many Westerners is the belief that civic action projects are always positive and relatively simple to execute. The idea that local populations must perceive such activities as beneficial is just not true. Military review units are ideally suited. Special Forces, foreign area officers, and soldiers working in civil affairs and psychological operations receive language and regional training. The level of that training varies depending on the region and on current requirements and priorities in Iraq and Afghanistan. It is common, however, for "specialists" on Africa to have no training on Africa and to have never deployed anywhere on the continent. Thus, even if regional specialists are available and we utilize them effectively, they may lack expertise.

---

4 "Special Forces to Serve at U.S. Embassies," *The Citizen* (Dar es Salaam. Tanzania), March 2006.

To make up for this, some military units use chaplains as culture specialists. Their commanders consider this a natural fit, given the close link between religion and culture. But while chaplains have an assigned role to advise commanders on religious matters in military operating environments – a role they have generally performed with great success in Iraq – having to deal with culture as a whole will create a dilemma for them: How do they segregate religion from culture? This is an all-but-impossible task.

Components of culture cannot be isolated from each other, and broader cultural analysis is not an area in which chaplains are trained. Advising on religious considerations in an AOR is also a vague doctrinal role and brings into question the extent to which chaplains should perform missions interacting with locals outside of military bases, since many might view chaplains as biased, dogmatic, or ethnocentric. This is ultimately a command decision, and the point here is simply that commanders need to be aware of potential negative effects from the use of chaplains as cultural advisors and liaison officers.

These nontraditional missions may have unintended consequences. For example, a senior U.S. military chaplain recently requested permission to enter Tanzania to meet with key national religious leaders. His intent was to "[develop] ways in which religion, [a component] that plays a critical role in international relations here in this region, can be used as a force for peace and cooperation." His justification for visiting Tanzania further stated, "We have also sent donations by way of others to make their way into Southern Sudan. We liaise with secular and religious nongovernmental organizations (NGOs) throughout our Area of Interest (AOI) to leverage more efficient and effective shared goals."[5] All aid, humanitarian or otherwise, has at least some political and even military significance, but Christian NGOs fund the Sudan People's Liberation Army outright. By using his military position to funnel aid to the Sudan, the chaplain was consciously or unconsciously pursuing a politico-religious agenda; he was circumventing controls put in place by the U.S. government to prevent such actions.

The U.S. system of governance includes the separation of church and state; thus, no government agency has a mandate to do religious work. Chaplains in the U.S. military, however, are something of an anomaly. Because they are paid by the government specifically to minister to soldiers, there is no disguising the fact that they are religious advocates. The ill-advised use of the word "crusade" by American military and political leaders to describe the war in Iraq might make the chaplain look, to Arab-Muslim eyes, like a crusader,

5 Personal correspondence of the author.

a Judeo-Christian jihadist ("crusade" in Arabic translates as harb al salibeya: a war of the cross, which can easily be translated as "Christian jihad").[6] In two cases I observed in Iraq, this was underscored by chaplains carrying weapons, an act of questionable legality that violates the tenets of common sense and reinforces impressions of interfaith warfare.

For these reasons, designating military chaplains, who are overwhelmingly Christian, as cultural experts and as the primary agents for cultural interaction might give American regional activities a religious tinge. This is not an indictment of chaplains, but a cautionary note about the potential liabilities inherent in using chaplains in expanded roles in some politico-religious contexts. Overall, using chaplains as cultural specialists and advisers underlines the failure of the military chain of command to understand the complexities of local culture. In turn, this highlights the need for methodologically analyzing and integrating cultural factors into military operations.

Lessons lost

Using its operations in East Africa as a case in point, it is evident that the U.S. military has not applied lessons learned in Iraq. Thus far, U.S. forces bound for East Africa have received no training on East African culture prior to deployment; instead, the Army trained them for Iraq and Afghanistan. While much of this training was undeniably good – it included convoy live-fires; prisoner handling; and study of the law of war, small-unit tactics, and IED-recognition techniques – it simply wasn't applicable to operations in East Africa. Consequently, U.S. forces in the region have often relied on the U.S. Embassy for basic assistance, both logistical and informational. This can lead to clashes with embassy personnel, who may see U.S. military forces new to a region as a drain on time and resources and as a potential source of embarrassment.

The lack of regional training and overall expertise also prevents U.S. forces from adequately integrating into foreign societies. They sometimes reside in luxury hotels and hire translators or "expeditors" to procure items in the local economy and to advise them on how to interact with locals. Sustained operations have involved the creation of luxurious "safe houses" in the

---

6   For instance, on 16 September 2001, during a televised speech and press conference, President Bush said, "This crusade, this war on terrorism is going to take awhile...." See Peter Ford, "Europe cringes at Bush 'crusade' against terrorists," *The Christian Science Monitor*, 19 September 2001, 12. During Desert Storm, the author observed a tank with "CRUSADER" painted on the barrel, and another example is the Army's Crusader artillery system.

wealthy expatriate communities of East Africa. Although this arrangement meets embassy guidelines for force protection and helps keep forces under some form of control through proximity, it doesn't provide the optimum setting in which to learn about a country.

If the U.S. military is going to conduct peacetime engagement activities, it must incorporate ever-changing socioeconomic, cultural, ethnic, and historical knowledge into operations planning and execution, and it must give its leaders access to information and specialists so they can make informed decisions. We must overcome dogmatic institutional prerogatives. We need mature, informed decisions influenced by feedback. We must build an institutional knowledge base that gives us flexibility and continuity.

One cannot understand the conditions that breed terrorism by observing them from the isolation of luxurious enclaves in capital cities during a 90-day stint of temporary duty. It takes years of training, and it takes command recognition that the mission is important.

12

On the Uses of Cultural Knowledge

Sheila Miyoshi Jager

Summary

The wide-spread recognition of the need for cultural knowledge in counterinsurgency has been noted and actively promoted recently by the Department of Defense (DoD). General David H. Petraeus, commanding general of the Multi-National Force Iraq (MNF-I), has been at the vanguard of these efforts. As the commander of the 101st Airborne Division in the initial invasion of Iraq in 2003, he later took responsibility for governing Mosul, Iraq's second largest city. Relying on his experiences in Mosul, General Petraeus is currently in charge of a major new counterinsurgency effort in Iraq.

In sharp stark contrast to former Secretary of Defense Donald Rumsfeld's heavy-handed approach to counterinsurgency which emphasized aggressive military tactics, the post-Rumsfeld Pentagon has advocated a "gentler" approach, emphasizing cultural knowledge and ethnographic intelligence as major components of its counterinsurgency doctrine. This "cultural turn" within DoD highlights efforts to understand adversary societies and to recruit "practitioners" of culture, notably anthropologists, to help in the war effort in both Iraq and Afghanistan. The recent focus on cultural knowledge in counterinsurgency operations and tactics is a welcome development insofar as it has allowed field commanders in Iraq and Afghanistan to reassess radically the failed operations and tactics in counterinsurgency in both these places. However, what has so far been absent from the discussion on cultural knowledge is the effort to link this new knowledge to formulating an overarching strategic framework. If cultural knowledge has helped U.S. forces to refocus their efforts to achieve better their operational and tactical goals, the question our political leaders should be asking is whether cultural knowledge can also help them to redefine a broader strategic framework for counterinsurgency.

The aim of this monograph is two-fold. First, it attempts to distinguish between the various "levels" of cultural knowledge and how they are used at various levels of warfare – strategy, operations, and tactics. Although not mutually exclusive, cultural knowledge informs these distinct levels in different ways. For example, the kinds of cultural knowledge that are required at the tactical level (e.g., the cultural knowledge of specific customs) is quite separate from the kinds of cultural knowledge that are required to formulate grand strategy and policy.

Second, the monograph attempts to explore how cultural knowledge might help to redefine an overarching strategy on counterinsurgency. While the military has been at the forefront of significant new and innovative thinking about operations and tactics, revising its old doctrines on the fly, America's political leaders have failed to provide the necessary strategic framework to guide counterinsurgency. The innovative insights about cultural knowledge adapted in operations and tactics by our military leaders have so far not yielded any comparable innovations from our political leaders. While the use of cultural knowledge is transforming military operations and tactics in significant and revolutionary ways, this same knowledge is not being adapted by our political leaders to help redefine a compelling new strategy for counterinsurgency.

The monograph concludes by suggesting four distinct ways in which cultural knowledge can work to help redefine an overarching strategic framework for counterinsurgency.

1.   Reconceptualizing the "war on terror" not as one war, but as many different wars.

2.   Focusing less on the moral distinctions between "us" and "them" – a major centerpiece of the Bush Doctrine – and more on the differences between "them."

3.   Building support and relationships among both friendly and adversary states by taking into account how other societies assess risks, define their security, and perceive threats.

4.   Building support for counterinsurgency among America's civilian leaders. Especially amid the domestic acrimony spawned by the Iraq War, inadequate coordination between military and nonmilitary power will severely hamper U.S. counterinsurgency capabilities. Cultural knowledge of both military and civilian institutions is therefore vital if the coordination between them is to be effective.

On the uses of cultural knowledge

Knowledge of the cultural terrain can be as important as, and sometimes even more important than, the knowledge of the geographical terrain. This observation acknowledges that the people are, in many respects, the decisive terrain, and that we must study that terrain in the same way that we have always studied the geographical terrain.

–General David H. Petraeus, Commanding General, Multi-National Force Iraq (MNF-I)[1]

Culture has become something of a buzzword among America's national security leaders. Faced with a brutal civil war and insurgency in Iraq, the many complex political and social issues confronted by U.S. military commanders on the ground have given rise to a new awareness that a cultural understanding of an adversary society is imperative if counterinsurgency is to succeed. Now embroiled in a counterinsurgency in Iraq with no clear end in sight, the broad outlines of what went wrong in Iraq – from insufficient post-war planning to de-Ba'thification and demilitarization of Iraqi society that led to the subsequent emergence of old tribal networks and ethnic and religious cleavages – have been traced to a glaring misunderstanding of Iraqi culture and society by American occupation planners and U.S. military forces. American occupation planners simply assumed that the civilian apparatus of the government would remain intact after the regime was decapitated by the military defeat. But in fact, "when the United States cut off the hydra's Ba'athist head, power reverted to its most basic and stable form – the tribe."[2] Without a firm understanding of the cultural dynamics of Iraqi society or the brutal legacy of colonialism and Sadaam's persecution of Iraq's Shi'ite and Kurdish population, American occupational forces in Iraq were basically working within a cultural and historical vacuum.

The new efforts to infuse cultural knowledge into U.S. military operations and training in Iraq have coincided with a broad shift within the Department of Defense (DoD), once the extent of the debacle in Iraq became more widely known. In July 2004, retired U.S. Army Major General Robert H. Scales, Jr., wrote an article for the Naval Institute's *Proceedings* magazine in which he disagreed with the commonly held assumption that

1 General David H. Petraeus, "Learning Counterinsurgency: Observations From Soldiering in Iraq," *Military Review, Special Edition Counterinsurgency Reader*, October 2006, p. 51.

2 Montgomery McFate, "The Military Utility of Understanding Adversary Culture," Journal of Defense Analysis, Issue 38, p. 44.

was prevalent within the Pentagon at the time – that success in war is best achieved by overwhelming force. Instead, he argued that the type of conflict we are currently waging in Iraq requires "an exceptional ability to understand people, their culture, and their motivations."[3] Since then, the widespread recognition of the need for cultural knowledge in counterinsurgency has been recognized and actively promoted by the Pentagon. General David H. Petraeus, commanding general of the Multi-National Force Iraq (MNF-I), who also boasts a Ph.D. from Princeton in International Relations, has been at the vanguard of these efforts. As the commander of the 101st Airborne Division in the initial invasion of Iraq in 2003, he later took responsibility for governing Mosul, Iraq's second largest city. Relying on his experiences in Mosul, General Petraeus is currently in charge of a major new counterinsurgency effort in Iraq. Desperate to stem the on-going violence, the Bush administration is pinning its hopes on General Petraeus and his advisors to fix the fiasco in Iraq.

In sharp stark contrast to then Secretary of Defense Donald Rumsfeld's heavy-handed approach to counterinsurgency which emphasized aggressive military tactics, the post-Rumsfeld Pentagon has advocated a "gentler" approach, emphasizing cultural knowledge and ethnographic intelligence as major components of its counterinsurgency doctrine. This "cultural turn" within DoD highlights efforts to understand adversary societies and to recruit "practitioners" of culture, notably anthropologists, to help in the war effort in both Iraq and Afghanistan. In February 2006, Petraeus invited an array of academics, human rights lawyers, journalists, and practitioners of counterinsurgency to Fort Leavenworth, Kansas, to vet a draft for a new counterinsurgency manual, Field Manual (FM) 3-24, which was published on December 15, 2006. Owing to its enormous popularity, however – with 1.5 million downloads the first month – it was recently republished by the University of Chicago Press with a forward by Sarah Sewell, a former DoD official who now teaches at the Kennedy School of Government at Harvard University.

While the focus on cultural knowledge in counterinsurgency operations and tactics is a welcome development insofar as it has allowed

---

3   Major General Robert H. Scales, "Culture-Centric Warfare," *Proceedings*, October 2004, p. 21. See also Scales, "Army Transformation: Implication for the Future," Testimony before the Armed Services Committee, Washington D.C., July 2004. See also "Counterinsurgency Reader: Special Edition," *Military Review*, October 2006. This manual complements the Army/ Marine Corps Field Manual (FM) 3-24 on counterinsurgency operations.

field commanders to reassess radically the failed operations and tactics in counterinsurgency, what so far has been absent from the discussion on cultural knowledge is the effort to link this new knowledge to formulating an overarching strategic framework. As Sarah Sewell has put it, "because counterinsurgency is predominately political, military doctrine should flow from a broader strategic framework. But our political leaders have so far been unable to provide a compelling one."[4] If cultural knowledge is now viewed as a major component of counterinsurgency operations and tactics on the ground, what can cultural knowledge teach us about strategy and policy? This question requires us to distinguish the various "levels" of cultural knowledge and how they are used at various levels of warfare – strategy, operations, and tactics. Although not mutually exclusive, cultural knowledge informs these distinct levels in different ways. For example, the kinds of cultural knowledge that are required at the tactical level (e.g., the cultural knowledge of specific customs like "do not spit in public," or "take off your shoes before entering a house," etc.) is quite separate from the kinds of cultural knowledge that are required to formulate grand strategy and policy (e.g., the cultural knowledge that influences such broad issues as how the legacy of Japanese imperialism has influenced contemporary Sino-Japanese relations).

However, within the current literature on culture and counterinsurgency, there has been a tendency to conflate the practical application of empirical cultural knowledge (as applied to operations and tactics) with the more abstract notions of cultural knowledge as they apply to the formulation of an overarching strategy and policy for counterinsurgency. The kinds of cultural knowledge that inform military operations and tactics on the ground – the "how-to" practical application of cultural and ethnographic knowledge – is very distinct from the forms of cultural knowledge that are needed to formulate national strategy and policy. However, although quite distinct, the uses of culture as they apply to all three levels are interrelated and must complement one another: a sound strategic framework based on a deep cultural and historical understanding of an adversary culture will necessarily give rise to sound operations and tactics necessary for waging a successful counterinsurgency.

Thus far, there has been a great deal of concern with the application of cultural knowledge on the battlefield and far less interest in how this knowledge might be applied to formulating an overarching strategic framework on counterinsurgency. Without a clear articulation of our strategic objectives,

---

4 Sarah Sewell, "Introduction to the University of Chicago Press Edition: A Radical Field Manual," *Counterinsurgency Field Manuel*, Chicago: University of Chicago, 2007, p. xl.

our political leaders have confused operational and strategic goals. Achieving stability in Iraq is an operational goal; it is not a strategic objective. Devising a broad strategy for counterinsurgency requires our political leaders to focus their attention beyond the counterinsurgency in Iraq and Afghanistan.

If cultural knowledge has helped U.S. forces refocus their efforts to achieve better their operational and tactical goals, the question our political leaders should be asking is whether cultural knowledge can also help them redefine a broader strategic framework for counterinsurgency. The answer to this question requires an examination of cultural knowledge and how it operates in different ways according to the different levels of war-making (strategy, operations, and tactics).

Cultural knowledge for strategy

What do we mean by cultural knowledge as applied to the level of grand strategy? How is it distinguished from the kinds of cultural knowledge needed to wage successful operations on the battlefield? Let us begin by using the definition of culture as articulated by the new National Cultures Initiative at the Department of National Security and Strategy at the U.S. Army War College (USAWC).

"Culture" is a difficult concept to grasp with any certainty, but a fundamental one defining and understanding the human condition. It is also an important dimension of policy and strategy, because it affects how people think and respond and thus how policy and strategy are formulated and implemented. We can consider culture as the way humans and societies assign meaning to the world around them and define their place in that world. It is manifested in languages, ideas, beliefs, customs, traditions, rituals, objects and images that are symbolic (therefore symbolic forms that represent and/or contain certain meanings) of the values, interests, perceptions, and biases of individuals and of the collective society.[5]

Largely in response to the setbacks in Iraq, the USAWC has introduced major new changes to its curricula which have sought to address the issue of culture directly. As part of the "cultural turn" within the DoD, new lessons on National Cultures in the standard Strategic Thinking course and a new series of Regional Studies courses were introduced into the curriculum in 2006-07. The aim of these courses is to teach students about the importance

---

5   Colonel Jiyul Kim, "The Analytical Framework," Course Reader for the Strategic Thinking Course, Academic Year 2008, Carlisle Barracks, PA: U.S. Army War College, p. 2.

of cultural awareness and understanding of "how other regions, nations, and societies view themselves and others" and the effect of this awareness on policy and strategy formulations and outcome. This is a significant shift away from the traditional focus on American interest and policy in foreign areas. Led in large part by Colonel Jiyul Kim, Director of Asian Studies at the USAWC, the Analytical Culture Framework, which serves as a master guide to these major new efforts and which he authored, lists six dimensions for the study of culture that form the intellectual framework for the new Strategic Thinking and Regional Studies courses. These dimensions are:

1. National identity
2. Political culture
3. Regional identity
4. Political system
5. Strategic culture, and
6. Globalization and culture.

A common theme that infuses all six dimensions is the critical place occupied by the study of history: every dimension of the framework must be appreciated as both a cumulative and revisionist process of not only the actual historical experience, but also memory of that history for memory often distorts history for contemporary purposes. Thus history serves two important functions, as agent and process that actually determines specific cultural forms (both tangible and intangible), and as an instrument of culture to be distorted and used for contemporary purposes (most often political).

These new curricular initiatives are significant in their attempt to link the understanding of foreign cultures at its most abstract level (national identity, political culture, strategic culture, etc.) with American strategy and policy: "We live in a world without the comfortable and simple dichotomy of the Cold War . . ." reads the National Cultures lesson. Greater cultural proficiency at the strategic level is imperative in working with the rising powers such as China and India, dealing with new partners and allies as well as new challenges with old allies and partners, responding to extremism in its many forms, learning to wage an effective counter-insurgency campaign, coping with increasing anti-Americanism, handling transnational threats and issues, and building coalitions across the regions and the world.[6] While this linkage between cultural knowledge and U.S. strategy appears to be new, culture figured prominently in America's post-World War II planning. The successful military

---

6 Colonel Jiyul Kim, "National Cultures," Course Reader for the Strategic Thinking Course, Academic Year 2008, Carlisle Barracks, PA: U.S. Army War College, 2007, p. 1.

occupation of Japan (1945-52) is a good example of how cultural knowledge informed America's long-term strategic objectives in Asia. The U.S. decision to preserve the Japanese imperial system and shield Emperor Hirohito from being tried as a war criminal (something that was fiercely opposed by Japan's neighbors and many political groups within the United States) allowed the American Occupation to rewrite a new role for the Japanese Emperor:

Hirohito was miraculously transformed from Japan's preeminent military leader who oversaw a brutal 15-year war against Asia and the United States to an innocent Japanese victim and political symbol duped by evil Japanese militarists. The surprising and rapid transition from Japanese militarism to Japanese democracy was made not through the imposition of American democratic values and norms, but by a not-so-subtle manipulation of Japanese cultural symbols and meanings, including a rather blatant manipulation of history.[7]

Applied to the level of strategy, cultural knowledge must therefore take into account the vital role of history and historical memory. Culture is not unchanging, nor does it entail a set of enduring values and/or ancient "patterns" of thought from which we can predict behavior. This is where the usage and understanding of culture as applied to the level of strategy differs significantly from the application of cultural knowledge at the operational and tactical levels. The uses of cultural knowledge in counterinsurgency operations emphasize the need for soldiers to understand the intricacies of customs, values, symbols, and traditions in order to be able to adapt and fight in a foreign society. It is hoped that this anthropological approach to war "will shed light on the grammar and logic of tribal warfare," and create the "conceptual

---

7 The post-war American occupation of Japan also possessed a great intangible quality that simply was never present in Iraq: it enjoyed virtually unquestioned legitimacy – moral as well as legal – in the eyes not merely of the victors, but all of Japan's neighbors who had been victimized during the course of Japan's brutal 15 years of war. There had been a formal surrender by the Japanese, and the American occupation and reconstruction had been endorsed by Emperor Hirohito. Furthermore, except for the military, the Japanese government remained intact at all levels, and, of course, the Japanese already had a tradition of democracy and civil society on which to draw (during the Taisho era, 1912-26). There were also no hostile or religious factions within the country. But mostly, planning for the occupation of Japan had begun already in 1942, giving planners ample time to think through many of the cultural and political issues that the American Occupation would confront after the surrender. See John Dower, "A Warning for History: Don't Expect Democracy in Iraq," Boston Review, February/March 2003. On the American occupation of Japan, see John Dower, *Embracing Defeat: Japan in the Wake of World War II*, New York: W. W. Norton, 1999.

weapons necessary to return fire."[8]

Against this definition of culture as an enduring "grammar" of values and customs rooted in a timeless tradition, cultural knowledge as applied to the level of strategy assumes that cultures are dynamic entities, not static categories. Hence, in formulating an overarching strategic framework for counterinsurgency, it is important to grasp not merely the cultural logic of, say, Sunni identity, including their values, customs, traditions, etc., but how Sunni extremists have invoked these traditional values, historical experiences, and belief-systems in the contemporary context to justify their extremist actions. Culture as applied to the level of strategy focuses on the issues of interpretation and reception. Cultural knowledge at this level thus requires a complex understanding of culture as a dynamic entity, an on-going process of negotiation between past and present. Far from reproducing the values and beliefs of a static and unchanging culture, extremist groups like Al-Qaeda have appropriated and reinterpreted Islamic texts, belief-systems, and traditions to justify their own radical ideology; in other words, they have used culture instrumentally. Cultural knowledge as applied to the level of strategy must be concerned with the dynamic understanding of culture and how different Islamic radicals emphasize different aspects of their historical past and traditions to legitimize their political actions and behavior in the present. Such knowledge becomes useful in formulating a grand strategy on counterinsurgency that, instead of lumping all Islamic radical enemies together, differentiates them according to their various "cultures" within radical Islam. To pry apart violent Islamic radicals, the United States has to become knowledgeable about these internal cultural cleavages "and be patient in exploiting them."[9] Cultural knowledge at the strategic level serves this purpose.

Cultural knowledge for operations and tactics

Cultural knowledge as applied to the level of operations and tactics is concerned with the practical application of this knowledge on the battlefield. In contrast to the dynamic understanding of culture and its usage at the level of strategy, culture at the operational and tactical levels is defined as a more

---

8 Patrick Porter, "Good Anthropology, Bad History: The Cultural Turn in Studying War," *Parameters*, Vol. xxxvi, No. 2, Summer 2007, p. 49.

9 Samantha Powers, "Our War on Terror," *New York Times Sunday Magazine*, July 29, 2007, p. 8. See also Ian Shapiro, *Containment: Rebuilding a Strategy Against Global Terror*, Princeton, NJ: Princeton University Press, 2007.

or less stable and static set of categories that include distinct belief systems, values, customs, and traditions that can be usefully applied to enhance the cultural awareness of American-led forces on the ground. It is primarily this understanding and usage of culture that have become prominent features of the counterinsurgency efforts in Iraq and Afghanistan.[10] Two major efforts in this regard are notable. As part of a new program to help address the short-comings in cultural knowledge by soldiers on the ground, the Foreign Military Studies Office (FMSO), a U.S. Training and Doctrine Command (TRADOC) organization that supports the Combined Arms Center at Fort Leavenworth, Kansas, is overseeing the creation of the Human Terrain System (HTS). According to its creators, this system is being specifically designed to address cultural awareness shortcomings at the operational and tactical levels by giving brigade commanders an organic capability to help understand and deal with "human terrain"—the social ethnographic, cultural and economic, and political elements of the people with whom the force is operating.[11] HTS is built upon seven components, or "pillars":

1.   Human terrain teams (HTTs)
2.   Reach-back research cells
3.   Subject-matter expert-networks
4.   A tool kit
5.   Techniques
6.   Human terrain information, and
7.   Specialized training.

Each HTT will be comprised of experienced cultural advisors familiar with the area in which the commander will be operating. The experts on the ground, these advisors will be in direct support of a brigade commander. All will have experience in organizing and conducting ethnographic research in a specific area of responsibility, and they will work in conjunction with other social science researchers. HTTs will be embedded in brigade combat teams, providing commanders with an organic ability to gather, process, and interpret

---

10 In addition to the practical application of cultural knowledge advocated in FM 3-24, the U.S. Army War College has adopted Richard D. Lewis's *When Cultures Collide: Managing Successfully Across Culture* (1999) in its curricula. This book also represents a "how-to" approach to cultural knowledge.

11 Jacob Kipp, Lester Grau, Karl Prinslow, and Captain Don Smith, "The Human Terrain System: A CORDS for the 21st Century," *Military Review*, September/October, 2006, p.10. "Human Terrain" Studies date back to Lieutenant Colonel Ralph Peter's work on "human terrain operations." See Ralph Peters, "The Human Terrain of Urban Operations," *Parameters*, Spring 2000, pp. 4-12.

relevant cultural data. In addition to maintaining the brigade's cultural data bases by gathering and updating data, HTTs will also conduct specific information research and analysis as tasked by the brigade commander.[12] These efforts represent the "how-to" practical application of cultural knowledge at the operational and tactical level. Designed specifically to teach cultural awareness as a battlefield skill, HTTs are also designed as data gathering systems for acquiring cultural knowledge for the purposes of providing new and incoming commanders and units with the "institutional memory" about the people and culture of their area of operation. In 2006, five HTTs deployed from Fort Leavenworth to Afghanistan and Iraq. If they prove successful, an HTT will eventually be assigned to each deployed brigade or regimental combat team.[13] Another central feature of the Human Terrain System is the emphasis on human relationships. "To be successful, you must understand the Iraqi perspective. Building trust, showing respect, cultivating relationships, building a team, and maintaining patience are all central features of the human terrain system which emphasize the power of people – friendship, trust, understanding – the most decisive factor in winning the war in Iraq."[14] The other significant product that has come out of the wars in Iraq and Afghanistan is the new counterinsurgency manual, FM 3-24. Released on December 15, 2006, FM 3-24 is the first U.S. Army manual dedicated exclusively to counterinsurgency in more than 20 years.[15] The 282-page document, like the HTS, highlights cultural knowledge and human relationships as central aspects for waging a successful counterinsurgency. These are highlighted in the first chapter under "Ideology and Narrative":

> Culture knowledge is essential to waging a successful counterinsurgency. American ideas of what is "normal" and "rational" are not universal. To the contrary, members of other societies often have different notions of rationality, appropriate behavior, levels of religious devotion, and norms concerning gender. Thus, what might appear abnormal or strange to an external observer may appear as self-evidently normal to a group member. For this reason, counterinsurgents – especially commanders,

---

12 Ibid., p. 12

13 For a good report and description of the first HTTs deployed to Iraq and Acfghanistan, see David Rohde, "Army Enlists Anthropoligy in War Zones," New York Times, October 5, 2007. www.nytimes.com/2007/10/05/world/asia/05afghan.html?_r=2&hp+slogin&pagewanted =print&oref=slogin.

14 Ibid., p. 13.

15 "Counterinsurgency," FM 3-24, at www.fas.org/irp/doddir/army/fm3-24.pdf. The FM mentions "culture" 88 times and "cultural" 90 times.

planners, and small-unit leaders – should strive to avoid imposing their ideals of normalcy on a foreign cultural problem.[16]

Chapter 3, "Intelligence in Counterinsurgency," defines terms including society, social structure, rules, and norms and social norms. It also emphasizes the importance of culture as a "web of meaning shared by members of a particular society or group within a society. Culture might be described as an operational code that is valid for an entire group of people.... [It] influences how people makes judgments about what is right and wrong."[17] Another section highlights identity, values, belief systems, and cultural forms. Listed under the cultural forms section are ideologies and narratives:

> The most important cultural form for counterinsurgents to understand is the narrative. A cultural narrative is a story recounted in the form of a casually linked set of events that explains an event in a group's history and expresses values, character, or self-identity of the group.
>
> Narratives are the means through which ideologies are absorbed by members of a society. ... By listening to narratives, counterinsurgents can identify a society's core values. Commanders should pay particular attention to cultural narrative of the HN (host nation) population pertaining to out-laws, revolutionary heroes, and historical resistance figures. Insurgents may use these narratives to mobilize the population.[18]

In chapter 5, "Executing Counterinsurgency Operations," the manual encourages the development of counternarratives "which provide a more compelling alternative to the insurgent ideology and narrative. Intimate cultural familiarity and knowledge of insurgent myths, narratives and culture are a prerequisite to accomplishing this."[19] One of the major innovations of FM 3-24 is its rejection of the notion that human behavior is motivated purely by rational self-interest. Instead, FM 3-24 proposes that culture informs individual actions, whether one society deems these actions "rational" or not. Culture, it insists, shapes the ways in which others perceive us and the world, and hence cultural knowledge of the adversary society must be a major component of counterinsurgency.

FM 3-24 has been described as "radical" and "revolutionary" by *Time*

---

16 Field Manual (FM) 3-24, "Counterinsurgency," Chicago: University of Chicago Press, 2007, p. 1-5.

17 Ibid., p. 3-8.

18 Ibid.

19 Ibid, p. 5-1.

magazine, and it has received rave reviews in the *New York Times*.[20] Understanding the cause for FM 3-24's enthusiastic reception is itself noteworthy, notes Sarah Sewell, "because it seems to point to the overwhelming feeling of a majority of Americans that the United States is adrift in the world with no foreign policy to guide it in Iraq and elsewhere."[21] Americans are "simply confused about the nation's strategic purpose in wake of September 11, 2001."[22] Once again, Americans are wrestling with a "disillusionment about politics and military power, and the debacle in Iraq has reinforced a familiar cynicism that risks disengaging Americans from their government and America from the rest of the world."[23] In an attempt to understand America's new role in the world and also to stem the growing disillusionment about politics at home, they have looked to FM 3-24 for answers: "The doctrine's most important insight is that even – perhaps especially – in counterinsurgency, America must align its ethical principles with the nation's strategic requirements."[24]

But in explaining what "fighting well" means, FM 3-24 raises profound moral and ethical questions about what counterinsurgency actually entails.

Anthropology and the uses of cultural knowledge

Nowhere have the questions raised by FM 3-24 been argued more passionately and more fiercely than among anthropologists for whom these issues have both deep personal and professional resonance. As experts on cultural knowledge, anthropologists in particular have been eagerly sought out by the military for recruitment into counterinsurgency in Iraq and Afghanistan. Reactions to these recruitment efforts, however, have been decidedly cool, if not downright hostile.

Once called the "hand maiden" to colonialism, anthropology had enjoyed a long and fruitful relationship with national security agencies like the Central Intelligence Agency (CIA) and DoD, but this relationship abruptly ended following the close of the Vietnam War. Today, largely due to the disciplines' ethical codes and also its tendency to look inward and its turn toward

---

20 Joe Klein, "Good General, Bad Mission," *Time*, January 12, 2007; and Samantha Powers, "Our War on Terror," *New York Times*, July 29, 2007.

21 Sewell, p. xl.

22 Ibid.

23 Ibid.

24 Ibid., p. xxii

postmodernism and critical self-reflection, anthropology remains a rather insular field which attracts few readers beyond its disciplinary boundaries (ask
anyone to name the latest ethnography they have read recently, and you get the
point). Furthermore, in sharp contrast to the other social sciences (namely,
political science and economics), anthropology remains the least engaged with
national security and policymaking agencies within the U.S. government. The
American Anthropological Association's (AAA) current "Statement of Professional Responsibility" states that "Anthropologists should undertake no secret
research or any research whose results cannot be freely derived and publicly
reported … no secret research, no secret reports or debriefings of any kind
should be agreed or given."[25]

It therefore comes as no surprise that FM 3-24 has been received
with scathing criticism by many anthropologists, but most notably by Roberto Gonzalez who has criticized the manual for is "numbingly banal" material which, he notes "does not reflect current anthropology theory" but reads
more like a "simplified introductory anthropology textbook."[26] But the more
serious matter of Gonzales' critique is what he sees as a dangerous trend in
the co-optation of cultural knowledge for military purposes.[27] These concerns
are shared by other notable anthropologists, namely David Price and Hugh
Gusterson, who are deeply troubled by signs that "connections between anthropologists, military counterinsurgency experts, and intelligence agencies
are multiplying and deepening."[28] They are also concerned by the implication

---

25 American Anthropological Association (AAA), "Statement on Ethics: Principles of Professional Responsibility," adopted May 1971 (as amended through November 1986), on-line at
www. aaanet.org/committees/ethics/ethcode.htm.

26 Roberto J. Gonzales, "Towards a Mercenary Anthropology?The New U.S. Army Counterinsurgency Manual FM 3-24 and the Military-Anthropological Complex," Anthropology Today,
Vol. 23, No. 3, June 2007, p. 15.

27 For a rebuttal of Gonzales' criticisms of FM 3-24 and anthropology's contributions to
the manual, see David Kilcullen, "Ethics, Politics, and Non-State Warfare: A Response to
Gonzales;" and Montgomery McFate, "Building Bridges or Burning Heretics: A Response to
Gonzales," in Anthropology Today, Vol. 23, No. 3, June 2007.

28 Gonzales, p. 17. See also Hugh Gusterson, "Anthropology and the Military—1968, 2003,
and Beyond?" Anthropology Today, Vol. 19, No. 3, pp. 25-26; and David Price, "Interlopers and
Invited Guests," Anthropology Today, Vol. 18, No. 6, 2002, pp. 16-21; "Lessons from Second
World War Anthropology," Anthropology Today, Vol. 18, No. 3, pp. 14-20; and David Price,
"American Anthropologists Stand Up Against Torture and the Occupation of Iraq," Counter-
punch, Vol. 20, November 2006, www.counterpunch.org/price11202006.html. The Diane
Rehm Show also recently hosted a program entitled "Anthropologists and War" on October 10,
2007. See drshow@wamu.org for more information.

of this relationship and what it means for anthropology's professional ethics. And they are concerned that when ethnographic work is performed clandestinely, it can endanger informants by putting them and their families at risk. But mostly, they wonder whether using cultural knowledge for covert military operations will threaten the disciplinary integrity of anthropology itself by creating "mercenary anthropology" in which cultural knowledge itself is used as a weapon.[29]

Largely as a result of these critiques, Gonzales and Kanhong Lin submitted two resolutions to the American Anthropological Association (AAA) in November 2006. One condemned torture and "the use of anthropological knowledge as an element of torture," while the other condemned the U.S. occupation of Iraq. If passed, these resolutions, "will send an unambiguous message to the military and intelligence agencies seeking to recruit anthropologists (as well as anthropologists working on their behalf), namely that AAA members oppose wars of aggression and will stand united against activities that might breach our professional ethics."[30] As Gonzales noted:

> Although academic resolutions are not likely to transform U.S. Government policies (much less the practices of contractors to the military) these do articulate a set of values and ethical concerns shared by many anthropologists. They could potentially extend and amplify dialogue among social scientists around issues of torture, collaboration with the military, and the potential abuse of social science in the "war on terror." Anthropologists may well inspire others to confront directly – and resist – the militarization of their discipline at this critical moment in the history of the social sciences.[31]

Although the resolutions in themselves are nonbinding, their effects on the profession have been chilling, especially for new anthropology Ph.Ds who are contemplating working for the military or participating in programs like HTS. Steve Fondacaro, head of the Human Terrain project, confided recently that since the HTS's inception in 2006, he had been able to hire only a handful of anthropologists. One of those recently hired "admitted that the assignment came with huge ethical risks. I do not want to get anybody killed, she said. ... I end up getting shunned at cocktail parties, she said. I see there could be misuse. But I just can't stand to sit back and watch these mistakes happen over

---

29 Price, p. 3.

30 Gonzales, p. 19.

31 Ibid.

and over as people get killed, and do nothing.[32] The important issues raised by Gonzales and the AAA about the relationship between ethics and ethnography, namely, that FM 3-24 does look quite like "a suspect marketing campaign for an inherently inhumane concept of war," also raises significant questions about the uses of civilians in military operations. How should civilians respond to a war they condemn as immoral yet which requires their expertise to save American lives? Since the military's mission is to execute the policies of our democratically elected officials, can Gonzales and other anthropologists really deny commanders in Iraq and Afghanistan the cultural knowledge they need to wage a war they were charged by their political leaders with fighting? Is it ethically more correct for them to retreat from the world and leave others to do the fighting? Is the moral response to cynicism about politics and military power to do nothing, or in Gonzales' case, to censure those who choose to do something?

These debates among anthropologists, although academic and insular, are nonetheless instructive because they bring attention to the much larger debate that FM 3-24 raises for all Americans. These entail significant questions about civil-military relations and 18 the uses of civilians in military operations. The major premise of FM 3-24 is that successful counterinsurgency will require the efforts of both the military and civilians. The manual states quite explicitly that the burdens of waging counterinsurgency must be shared equally, and that it will require the efforts of the entire American population:

> Military forces can perform civilian tasks but often not as well as the civilian agencies with people trained in those skills. Further, military forces performing civilian tasks are not performing military tasks. Diverting from those tasks should be a temporary measure, one taken to address urgent circumstances.... The nature of the conflict and its focus on the populace make military and civilian unity a critical aspect of COIN operation.[33]

FM 3-24 asks civilian actors and agencies to be centrally engaged in the field alongside combat forces and share the risks of counterinsurgency equally with the military. As Sewell puts it, "it stresses the importance of effectively employing nonmilitary resources as power to share the burdens of a long-term, difficult, and morally questionable war. It tells Americans that if we fight these wars and if we wish to succeed with any approximation of honor, counterin-

---

32 George Packer, "Knowing the Enemy: Can Social Scientists Redefine the 'War on Terror'?" New Yorker, December 18, 2006, p. 6.

33 FM 3-24, pp. 2-9, 2-10.

surgency will demand more than we are accustomed to giving."[34] Ultimately, however, the demands of counterinsurgency may be too great for the American public to bear, not because of the significant costs and commitments involved, but because the ethical and moral dilemmas posed by counterinsurgency may drive Americans, like Gonzales, to retreat from the world and leave the fighting to the military.

Cultural knowledge for national strategy and policy

The greater challenge of counterinsurgency, however, lies not at the operational level but at the strategic one. While the military has been at the forefront of significant new and innovative thinking about operations and tactics, revising its old doctrines on the fly, America's political leaders have so far failed to provide the necessary strategic framework to guide counterinsurgency. The innovative insights about cultural knowledge adapted in operations and tactics by our military leaders have so far not yielded any comparable innovations from our political leaders. While culture is transforming the military in significant and revolutionary new ways, it seems to have had little impact on defining overall U.S. strategic goals. Now that U.S. Armed Forces are in Iraq, America's political leaders are consumed by how to get them out of it. Without an overarching strategy on counterinsurgency, our political leaders are focused on achieving short-term goals rather than long-term strategic objectives. Furthermore, the insights gleaned from cultural knowledge on operations and tactics are not being adapted by our political leaders to help redefine a compelling new strategy for counterinsurgency.

What is needed from our political leaders is an overarching strategic framework for counterinsurgency informed by culture. Internationally, the pursuit of regime change and radical visions of transforming the Middle East that were a primary tenet of the Bush Doctrine have proven costly. They have created instability in the region and resulted in the overextension of U.S. military power. President George W. Bush's "forward strategy of freedom" so far has failed to produce positive results in large part because it has advocated freedom without taking into account how that freedom would be received by other cultures.[35] The Bush "revolution" was about the imposition of American values, not about laying the groundwork for creating the necessary condi-

34 Sewell, p. xxxix.

35 "Forward Strategy of Freedom" was a phrase included in President Bush's remarks at the 20th anniversary of the National Endowment for Democracy, U.S. Chamber of Commerce, Washington, D.C., November 6, 2003.

tions for their reception.[36] Moreover, by creating a rigid line between "us" and "them," the Bush Doctrine lumped like and unlike foes together.[37] Unable to distinguish America's enemies abroad, the Bush administration treated all "terrorists" as a monolithic enemy (in Iraq and elsewhere). But this is precisely what George Kennan, who was a very good student of culture himself, had warned America's Cold War leaders against. Communism, he argued was not a monolith, and policymakers ought to be emphasizing and exploiting the differences among communists (as former President Richard Nixon did when he went to China in 1972). By failing to exploit the cultural distinctions and inherent tensions among our enemies, we have indirectly empowered them. "What these groups want" argued Hilary Benn, the British secretary of state for international development, "is to force their individual and narrow values on others, without dialogue, without debate, through violence. And by letting them feel part of something bigger, we give them strength."[38] Dissecting the calamities of the last 6 years of American foreign policy has become something of a sport, only because it has become all too easy to criticize. But as Samantha Powers warns, "it does not itself improve our approach to combating terrorist threats that do in fact loom – larger, in fact because of Bush's mistakes."[39] The challenge is to learn from these mistakes. What the failures of the Bush Doctrine have made abundantly clear is that cultural knowledge must be an important dimension of policy and strategy because it influences the way people think and respond and thus how policy and strategy are formulated and implemented.

---

36 The current U.S. National Security Strategy (NSS) is referred to as the "Bush Doctrine." Developed in response to circumstances confronting the United States and its allies in the wake of 9/11, it provides a stated use-of-force policy that addresses the requirement for offensive action to prevent threats from materializing on American shores. Lieber and Lieber (2007) have identified four key themes of the Bush Doctrine that have generated controversy. First, it calls for preemptive military action against hostile states and terrorist groups seeking to develop weapons of mass destruction (WMD). Second, it advocates that the United States will not allow its global military strength to be challenged by any foreign power. Third, it expresses a commitment to multilateral international cooperation, but makes clear that the United States "will not hesitate to act alone, if necessary" to defend national interests and security. Fourth, it proclaims the goal of spreading democracy and human rights around the globe, expecially in the Muslim world. See Keir A. Lieber and Robert J. Lieber, "The Bush National Security Strategy," usinfo.state.gov/ journals/itps/1202/ijpe/pj7-4lieber.htm.

37 See Ian Shapiro, *Containment: Rebuilding a Strategy Against Global Terror*, Princeton: Princeton University Press, 2007.

38 Quoted in Samantha Powers, "Our War on Terror," *New York Times*, July 29, 2007, p. 4.

39 Ibid.

How would cultural knowledge work to redefine a new policy and strategy for counterinsurgency? First, we could begin by reconceptualizing the "war on terror" not as one war but as many different wars. This means fighting terrorist groups and networks, even transnational ones like Al-Qaeda, as separate but related conflicts. This in turn implies flexibility and adapting our military operations and tactics to meet the distinct challenges of our enemies.

Second, a related aspect of this strategy would be to focus less on the moral distinctions between "us" and "them" – a major centerpiece of the Bush Doctrine – and more on the differences between "them."[40] This implies separating terrorist groups (as distinct social, cultural, and political entities) and also recognizing that although all of them hate America, they might hate each other even more. The more we learn to recognize and exploit the cultural differences among these terrorist groups, the better we will be able to isolate and defeat them. Of course, any effective campaign against terrorism must include political, economic, military, and paramilitary efforts along with cultural efforts. Stabilization is obviously a major strategic objective of counterinsurgency. But the ability to neutralize terrorist groups by playing up their differences, thus containing them by forcing them back into a local criminal or even political box, requires cultural knowledge. In a related context, by lumping North Korea, Iran, and Iraq together as one "axis of evil," instead of dealing with North Korea as a distinct cultural and political entity with its own history and grievances, the Bush administration got locked into an hostile, unproductive, and stubborn policy approach that went nowhere. It was only after the debacle in Iraq became fully apparent that the Bush administration finally backed down, but by then North Korea already had a nuclear bomb.

Third, antiterrorism efforts must also include building support and relationships among both friendly and adversarial states. It can be no longer a question of "you are either with us or against us," because counterinsurgency

---

40 A variation of the "us" verses "them" strategic outlook is the recent attempt by the Bush administration to create a new strategic alignment in the Middle East by separating "reformers" and "extremists," with Sunni states the centers of moderation and Iran, Syria and Hezbollah on the other side of that divide. This new policy for containing Iran is complicating the Bush administration's strategy for winning the war in Iraq. By insisting once again on imposing divisions from the outside instead of working to exploit the natural divisions within extremist groups, the Bush administration clumps our adversaries into one group, thereby empowering them. For a good overview of the new policy on Iran, see Seymour M. Hersh, "The Redirection: Is the Administration's New Policy Benefiting Our Enemies in the War on Terrorism?" *The New Yorker*, March 5, 2007.

requires too much work for the United States to go at it alone. This in turn implies both flexibility and deference in how U.S. strategic objectives for counterinsurgency are defined and executed.[41] Cultural knowledge of how other societies assess risks, define their security, and perceive threats all serve to underscore that cultural knowledge is an important dimension of how the United States must go about winning allies in the global war(s) on terror. Americans are not good at conceptualizing how other societies perceive the world and, in particular, how other societies perceive us. FM 3-24 represents a good starting point of how the United States must learn to get inside the minds of its adversaries.

Finally, the role of counterinsurgency and its relationship to U.S. national security must be explained to the American people. America's politicians must build support for counterinsurgency among America's civilian leaders. Especially amid the domestic acrimony spawned by the Iraq War, the inadequate coordination between military and nonmilitary power will severely hamper the kinds of U.S. counterinsurgency capabilities for which FM 3-24 has called.

To this end, cultural knowledge of both military and civilian institutions is vital if the coordination between them is to be effective. In particular, cultural knowledge of the military, its institutional values, traditions, historical role in society, and how it operates must be explained to the American public. More than the damage that Abu Ghraib did to America's image abroad, the scandal and its poor handling tarnished the military's image at home. Lingering moral doubts about the uses of counterinsurgency capabilities (like those expressed by anthropologists) in military operations may even provoke isolationist sentiments among the American public, leading to an unsteady retreat from abroad. As Sewell notes, "the very word counterinsurgency has become associated so closely with Iraq and a strategy of regime change that civil servants were loathe to consider themselves part of a U.S. counterinsurgency effort."[42] Explaining to the American public why counterinsurgency operations are important, and coordinating these efforts between military and civilian agencies to build a national consensus must be part of an overarching strategic framework for counterinsurgency in the post-September 11, 2001 (9/11) world.

But counterinsurgency is just one of the many challenges to U.S.

---

41 Sewell, p. xliii.

42 Ibid., p. xli.

security in the 21st century. Nuclear proliferation is another major challenge, as is the rise of China. All these challenges will require our political leaders to provide a new strategic vision for U.S. security. Already many scholars and practitioners have begun to interpret events like the U.S.-China standoff over a downed spy plane in 2001 or escalating tensions between Japan and China through the lens of national identity and culture.[43] These trends, like those already going on in the military, have profound implications for U.S. foreign policy. Armed with cultural knowledge, the United States will be better 24 able to restrain our adversaries through engagement by using shrewd diplomacy to dampen the strategic competition with China, Iran, and other potential rivals. A foreign policy guided by a deep understanding of the forces of nationalism, identity, and collective memory is a powerful tool to shape and mold adversarial behavior.

These forces, unwittingly unleashed by the Bush administration in the aftermath of the U.S. invasion of Iraq in 2003, now threaten the integrity of the Iraqi nation and have led to our current quagmire there. Although it may too late to save Iraq, it is not too late to apply the lessons that we have learned there to deal with other troubled spots in the world, namely North Korea, Iran, and China. If cultural knowledge has been able to reverse some of the operational and tactical blunders set forth by Rumsfeld's Pentagon, perhaps it not too late for culture to also rescue the United States from the strategic failures of the Bush Doctrine.

43   See Sheila Miyoshi Jager, *The Politics of Identity: History, Nationalism, and the Prospect for Peace in Post-Cold War Asia*, Carlisle Barracks, PA: Strategic Studies Institute, U.S. Army War College, April 2007. See also Graham E. Fuller, "America's Uncomfortable Relationship with Nationalism," Policy Analysis Brief, Muscatine, IA: The Stanley Foundation, July 2006; Thomas Berger, "Set for Stability? Prospect for Conflict and Cooperation in East Asia," Review of International Studies, Vol. 26, 2000, p. 420; Thomas Berger, *Cultures of Antimilitarism: National Security in Germany and Japan*, Baltimore: Johns Hopkins University Press, 1998; "Set for Stability? The Ideational Basis for Conflict in East Asia," *Review of International Studies*, Spring 2000; "The Construction of Antagonism: The History Problem in Japan's Foreign Relations," in G. John Ikenberry and Takashi Inoguchi, eds., *Reinventing the Alliance: U.S.-Japan Security Partnership in an Era of Change*, New York: Palgrave Macmillan, 2003; Peter Katzenstein, *Cultural Norms and National Security: Police and Military in Postwar Japan*, Ithaca, NY: Cornell University Press, 1996; Peter Katzenstein, ed., *The Cultures of National Security: Norms and Identity in World Politics*. See also Gerrit W. Gong, ed., *Remembering and Forgetting: The Legacy of War and Peace in East Asia*, Washington, DC: The Center for Strategic and International Studies, 1996; G. John Ikenberry and Michael Mastanduno, eds., *International Relations Theory and the Asia Pacific*, New York: Columbia University Press, 2003; and G. John Ikenberry and Takashi Inoguchi, eds., *Reinventing the Alliance: U.S.-Japan Security Partnership in an Era of Change*, New York, Palgrave Macmillan, 2003.

13

Female Suicide Bombers

Debra D. Zedalis

I have to tell the world that if they do not defend us, then we have to defend ourselves with the only thing we have, our bodies. Our bodies are the only fighting means at our disposal.

—Hiba, 28-year old, mother of five, suicide bomber trainee

Suicide bombers are today's weapon of choice. An action that was once so surprising, horrific, and terrifying has now become the daily fare of the nightly news. "From Jerusalem to Jakarta and from Bali to Baghdad, the suicide bomber is clearly the weapon of choice for international terrorists."[1] The raw number of suicide attacks is climbing;[2] suicide bombs are now used by 17 terror organizations in 14 countries.[3] "In terms of casualties, suicide attacks are the most efficient form of terrorism. From 1980 to 2001, suicide attacks accounted for 3 percent of terrorist incidents but caused half of the total deaths due to terrorism — even if one excludes the unusually large number of fatalities of 9/11."[4] Into this boiling cauldron of terror, a new element has been added — women as suicide bombers. The success of suicide bombers considerably depends upon surprise and accessibility to targets. Both of these requirements have been met by using women. The recent spate of female suicide bombers in different venues, different countries, and different terrorist organizations warrants careful study of this strategic weapon.

1   Don Van Natta, Jr., "Big Bang Theory: The Terror Industry Fields Its Ultimate Weapon," *New York Times*, August 24, 2003, sec. 4, p. 1.

2   Robert A. Pape, "Dying to Kill Us," *New York Times*, September 23, 2003, sec. 1A, p. 19.

3   Yoram Schweitzer, "Suicide Bombings: The Ultimate Weapon?" August 7, 2001. http://www.ict.org.il/articles/articledet.cfm?articleid=373.

4   Pape, sec. 1A, p. 19.

Prior to September 11, 2001, the use of suicide bombers was not seen by the U.S. public as a major threat; however, this weapon now evokes a visceral response. The Department of Homeland Security (DHS) Alert 03-025 provided warnings of male and female suicide bombers,[5] and New York City was on heightened alert in December 2003 over fears of an attack by a female suicide bomber.[6] Jessica Stern, a terrorism expert, recently criticized DHS procedures and warned of the dangers of ignoring female terrorists: "The official profile of a typical terrorist – developed by the DHS to scrutinize visa applicants and resident aliens – applies only to men. Under a program put in place after September 11, 2001, males aged between 16-45 are subject to special scrutiny; women are not."[7] Terrorists seek out vulnerabilities in the enemy government's countermeasures, so lack of scrutiny of women entering the United States could encourage Al-Qaeda to use them.[8]

This paper reviews the history of female suicide bombers, focuses on bomber characteristics, analyzes recent attacks, and predicts future trends within a strategic assessment of future female suicide bombings. According to a terrorist expert, "This research is particularly demanding as there are not enough reports with descriptions of the cases, there is not enough documentation on how these women are treated in terrorist groups, there are not enough testimonies, and it is difficult to profile these murderers."[9] Most chilling, however, is that those who could tell us their stories—those who went through with this death act—cannot share their thoughts and motivation.

History of suicide bombing

"A long view of history reveals that suicide terrorism existed as early as the 11th century. The Assassins (Ismalis-Nizari), Muslim fighters, adopted suicide terrorism as a strategy to advance the cause of Islam. These perpetrators perceived their deaths as acts of martyrdom for the glory of

---

5   Department of Homeland Security, Homeland Security Advisory System Increases to National Level ORANGE, Alert 03- 025, Washington, DC: U.S. Department of Homeland Security, May 20, 2003, p. 2.

6   "Terror Attack Fears," Sunday Mail (SA), December 21, 2003, sec. Foreign, p. 35

7   Jessica Stern, "When Bombers are Women," *Washington Post*, December 18, 2003, sec. 1A, p. 35.

8   Ibid.

9   Clara Beyler, "Messengers of Death, Female Suicide Bombers," February 12, 2003. http://www.ict.org.il/articles/articledet.cfm?articleid=471.

God."[10] Further, The Assassins' aim, like that of Islamist extremists today, was to spread a "pure" version of Islam. Like contemporary suicide bombers, they considered their own lives to be sacrificial offerings. Unlike today's suicide bombers, The Assassins murdered particular individuals rather than randomly targeting people whose only crime is to be in the wrong place at the wrong time.[11]

Female suicide bombers are relatively new. Their first known attack came in 1985 when a 16-yearold girl, Khyadali Sana, drove a truck into an Israeli Defense Force convoy and killed two soldiers. Since then, women have driven bomb-laden vehicles, carried bomber "bags," and strapped massive explosives and metal implements on their bodies in Lebanon, Sri Lanka, Chechnya, Israel, and Turkey. Terrorist groups which have publicized their use of females include the Syrian Socialist National Party (SSNP/PPS), the Liberation Tigers of Tamil Eelam (LTTE), the Kurdistan Workers Party (PKK), Chechen rebels, Al Aqsa Martyrs, Palestinian Islamic Jihad (PIJ), and, most recently, Hamas.

There are many "firsts" in this listing of organized feminine terror. While the SSNP has the distinction of deploying the first female suicide bomber, the "LTTE became the world's foremost suicide bombers and proved the tactic to be so unnerving and effective that their methods and killing innovations were studied and copied, most notably in the Middle East."[12] The LTTE has committed the most attacks, close to 200, using women bombers in 30-40 percent. The largest number killed (170) was in Moscow in October 2002 when Chechen rebels (including a high percentage of women) held hostages in the Theater Center, and the police killed 129 captives and 41 rebels in a futile rescue effort. During the last 2 years, Palestinian suicide bombers have carried out the largest number of attacks.

Not all bombers are known nor are all the data known to be correct. The youngest appears to be either Khyadali Sana (SSNP/PPS, 1985), 16, or Laila Kaplan, (PKK, 1996), 17; the oldest was Shagir Karima Mahmud (SSNP/PPS, 1987), 37. The first LTTE bomber, Dhanu, successfully killed Prime Minister Rajiv Gandhi in May 1991. The first female PKK suicide bombing (June 1996), which may also be the first instance of an apparently pregnant bomber, killed six Turkish soldiers; the bomber's name is unknown.

10 Ehud Sprinzak, "Rational Fanatics," *Foreign Policy*, September-October 2000, p. 68.

11 Jessica Stern, *Terror in the Name of God*, New York: Harper Collins, 2003, p. xxiii.

12 Amy Waldman, "Masters of Suicide Bombing: Tamil Guerrillas of Sri Lanka," *New York Times*, January 14, 2003, sec. A, p. 1.

The first Russian "Black Widow" or saliheen, Hawa Barayev, acted on behalf of the Chechen rebels in June 2000 and killed 27 Russian Special Forces soldiers. In January 2002, the first istish-hadiyat (female martyr) in Israel, representing the Al Aqsa Martyrs' Brigade, was Wafa Idris, a paramedic who detonated a 22-pound body bomb filled with nails and metal objects in a shopping district. Wafa killed an 81-year-old man and injured more than 100.

The first PIJ bomber was a 19-year-old student, Hiba Daraghmeh, who detonated a bomb in a shopping mall, killing three people. The second PIJ bomber, 29-year-old lawyer Hanadi Jaradat, also received much publicity as she strolled into a highly frequented restaurant in October 2003 and killed 21 Israeli and Arab men, women, and children. The most recent "first" was the first female Hamas bomber, 22-year-old Reem al-Reyashi, who, on January 14, 2004, killed four Israeli soldiers at a checkpoint. Of particular note is that Reem was also a mother who left behind a husband, a 3-year-old son, and a 1-year-old daughter. A "first" yet to strike is the first female suicide bomber representing Al-Qaeda (See Tables 1 and 2. Table 1 lists successful female suicide bombers. Table 2 shows those who were unsuccessful for some reason or who are trainees.)

Suicide bombers

A suicide bomber, someone willing to die for a cause, is puzzling. What is a suicide bomber? Why use suicide bombers? Who becomes a suicide bomber? Most importantly, why would one voluntarily die in order to kill other innocent people?

What is a suicide bomber?

The Institute for Counter-Terrorism (ICT) defined suicide bombing as an "operational method in which the very act of the attack is dependent upon the death of the perpetrator. The terrorist is fully aware that if she/he does not kill her/himself, the planned attack will not be implemented."[13] "Suicide bomber" is an emotionally-laden term. Some describe these individuals as "homicide bombers" or "suicide terrorists" to emphasize the murder and terror brought about by this act; others deem these individuals as "martyrs" who have died for their faith.

---

13 Boaz Ganor, "The First Iraqi Suicide Bombing. A Hint of Things to Come?" March 30, 2003. http://www.ict.org.il/articles/articledet.cfm?articleid-477.

Table 1. Successful Suicide Bombers.

Name: Andaleeb Takafka
Date: 12-Apr-02
Place: Israel
Age: 20
Status: Seamstress
Organization: Al Aqsa Martyrs Bde
Killed: 6
Wounded: 104
Notes: Fourth female suicide bomber. Detonated bag full of explosives at a bus stop. From Bethlehem.

Name: Aayal al-Akhras
Date: 29-Mar-02
Place: Israel
Age: 18
Organization: Al Aqsa Martyrs Bde
Killed: 2
Wounded: 25
Notes: Killed 2 in Jerusalem supermarket. Taped a martyr statement. Engaged. Came from Dehaisa Refugee Camp near Bethlehem. Cousins killed/wounded. Third female suicide bomber in Israel. Recently saw neighbor killed in his home while playing with his child.

Name: Dareen Abu Aysheh
Date: 27-Feb-02
Place: Israel
Age: 21
Status: Student
Organization: Al Aqsa Martyrs Bde
Killed: 0
Wounded: 4
Notes: Blew herself up at roadblock. Student at Al-Najah university. Came from village on West Bank. Went to Hamas to volunteer but they turned her down. Single; religious. Second female suicide bomber in Israel. Vowed revenge for Israeli troops shooting pregnant women at checkpoint. Cousin killed himself one month earlier.

Name: Wafa Idris
Date: 27-Jan-02
Place: Israel
Age: 28
Status: Paramedic
Organization: Al Aqsa Martyrs Bde
Killed: 1
Wounded: >100
Notes: First female suicide bomber in Israel. Detonated bomb in shopping district. Divorced, no children. Lived in Amari Refugee Camp. 22-lb. bomb with nails and metal objects. 3 brothers in Fatah. Treated wounded; hit 3 times by Israeli rubber bullets.

Name: Hawa Barayev
Date: 9-Jun-00
Place: Chechnya
Organization: Chechens
Killed: 27
Wounded: 0
Notes: Drove into building housing Russian Special Forces. Man accompanied her. First "black widow."

Name: Zulikhan Elikhadzhiyeva
Date: 5-Jul-03
Place: Russia
Age: 20
Organization: Chechens
Killed: 15
Wounded: 60
Notes: Two female suicide bombers at Moscow rock concert. Zulikhan was from Chechen village; brother is Wahhabi leader. Family home not destroyed; no close relatives killed. Some say brother kidnapped her.

Name: Unknown
Date: 6-Jun-03
Place: Russia
Organization: Chechens
Killed: 17
Wounded: 15
Notes: Flagged down bus carrying military personnel to air base. Third attack by female in 3 weeks. Officials believe rebels influenced by Islamic militants.

Name: Unknown
Date: 14-May-03
Place: Russia
Organization: Chechens
Killed: 16
Wounded:
Notes: 2 bombers. Religious festival. Tried to assassinate Russian leader, Akmud Kudyvor. Killed bodyguard.

Name: Unknown
Date: 1-May-03
Place: Russia
Organization: Chechens
Killed: 59
Wounded:
Notes: One woman and two men carried out truck bomb suicide attack. Attack at govt. complex in northern Chechnya.

Name: Reem al-Reyashi
Date: 14-Jan-04
Place: Israel
Age: 22
Status:
Organization: Hamas and Al Aqsa Martyrs Bde

Table 1 (continued)

Killed: 4
Wounded: 7
Notes: First Hamas suicide bomber; first mother in Israel. Requested medical help, blew up at inspection checkpoint. Made video (anti-Zionist). Left behind 3-year-old son and 1-year-old daughter. 10-lb. bomb w/ball bearings and screws. Came from middle-class family in Gaza City. Husband had no knowledge of her plans. Al Aqsa also claimed.

Name: Unknown
Date: Dec-99
Place: Sri Lanka
Organization: Liberation Tigers of Tamil Eelam (LTTE)
Killed:
Wounded:
Notes: Attempted to assassinate Sri Lankan President Chandrika Kumaratunga.

Name: Dhanu
Date: 20-May-91
Place: Sri Lanka
Organization: LTTE
Killed:
Wounded:
Notes: Killed Prime Minister Rajiv Gandhi

Name: Hanadi Jaradat
Date: 4-Oct-03
Place: Israel
Age: 29
Status: Lawyer
Organization: Palestinian Islamic Jihad (PIJ)
Killed: 21
Wounded: >12
Notes: At Maxims' Restaurant.

Name: Hiba Daraghmeh
Date: 1-May-03
Place: Israel
Age: 19
Status: Student
Organization: PIJ
Killed: 3
Notes: First Islamic Jihad; fifth female. Attack in mall.

Name: Ozen Fatima
Date: 17-Nov-98
Place: Turkey
Organization: Kurdistan Workers Party (PKK)
Killed: 0
Wounded: 6
Notes: Bomb strapped to body; missed target of military convoy.

Name: Otas Gular
Date: 29-Oct-96
Place: Turkey
Age: 29
Organization: PKK
Killed: 2
Wounded: 1
Notes: Killed policemen. Dressed as pregnant woman; accompanied by another group member. Third bombing by PKK females appearing to be pregnant. Ocalan had urged his troops to imitate Hamas by becoming human bombs.

Name: Laila Kaplan
Date: 25-Oct-96
Place: Turkey
Age: 17
Organization: PKK
Killed: 5
Wounded: 12
Notes: Attacked police headquarters as a pregnant woman.

Name: Unknown
Date: 30-Jun-96
Place: Turkey
Organization: PKK
Killed: 6
Wounded: 30
Notes: First female PKK bomber. Turkish soldiers killed. Bomb strapped to her stomach as if she were pregnant.

Name: Shagir Karima Mahmud
Date: 14-Nov-87
Place: Lebanon
Age: 37
Organization: Syrian Socialist National Party (SSNP/PPS)
Killed: 7
Wounded: 20
Notes: Carried explosive charge hidden in bag into hospital.

Name: Sahyouni Soraya
Date: 11-Nov-87
Place: Lebanon
Age: 20
Organization: SSNP/PPS
Killed: 6
Wounded: 73
Notes: Carried suitcase with explosives into airport. Case exploded by remote control.

Name: Al Taher Hamidah
Date: 26-Nov-85
Place: Lebanon
Age: 17
Organization: SSNP/PPS
Killed:
Wounded:
Notes: Drove car into SLA checkpoint; 100 kg of explosives.

Name: Khalerdin Miriam

Table 1 (continued)

Date: 11-Sep-85
Place: Lebanon
Age: 18
Organization: SSNP/PPS
Killed: 0
Wounded: 2
Notes: Attacked SLA checkpoint.

Name: Norma Abu Hassan
Date: 17-Jul-85
Place: Lebanon
Age: 26
Organization: SSNP/PPS
Killed: 0
Wounded: 7
Notes: Targeted Lebanese agents. Blew herself up when she saw soldiers searching for her.

Name: Kharib Ibtisam
Date: 9-Jul-85
Place: Lebanon
Age: 28
Organization: SSNP/PPS
Killed: 0
Wounded: 2 to 6
Notes: Attacked SLA post. Left videotape: "to kill as many Jews and their assistants" as she could.

Name: Khyadali Sana
Date: 9-Apr-85
Place: Lebanon
Age: 16
Organization: SSNP/PPS
Killed: 2
Wounded: 2
Notes: Drove car into IDF convoy. Soldiers injured/killed. Reason given: "avenge the oppressive enemy."

Name: Luisa Gazueva
Date: 29-Nov-01
Place: Chechnya
Age: Late 20s
Killed: 2
Wounded: 2
Notes: Attempted to kill Cdr Gaidar Gadzhiev. Young widow. Rebels did not claim the attack.

Name: Unknown
Date: 9-Dec-03
Place: Russia
Killed: 6
Wounded: 14
Notes: Outside the National Hotel in Moscow. Belt packed with 2.5 kg of plastic explosives and ball bearings – similar to belt from rock concert although detonating device was of lower quality.

Name: Unknown
Date: 1-Oct-02
Place: Russia
Killed: 170
Wounded:
Notes: Theater Center. 129 captives and 41 rebels killed.

Name: Rusen Tabanci
Date: 5-Jul-99
Place: Turkey
Age: 19
Killed: 0
Wounded: 17
Notes: Flashed "V" for victory and detonated bomb on body.

Name: Esma Yurdakul
Date: 27-Mar-99
Place: Turkey
Age: 21
Killed: 0
Wounded: 10

Name: Unknown
Date: 4-Mar-99
Place: Turkey
Killed: 0
Wounded: 4
Notes: In city square; bomb may have gone off prematurely as suspected target was police station.

Name: Unknown
Date: 24-Dec-98
Place: Turkey
Killed: 1
Wounded: 22
Notes: Outside Army barracks; killed herself and passerby.

Name: Unknown
Date: 1-Dec-98
Place: Turkey
Killed: 0
Wounded: 14
Notes: Kurdish woman outside supermarket frequented by Turkish soldiers.

Table 2. Unsuccesful Suicide Bombers.

Female suicide bombers who backed out before event, were unsuccessful and caught, or are currently in training. Those listed as trainees are identified by code name.

Name: Tawriya Hamamra
Date: May-02
Place: Israel
Age: 25
Status: Dressmaker/florist
Organization: Al Aqsa Martyrs Bde
Notes: From village near Jenin in the West Bank. Said reasons for attack were personal, not political. Received

Table 2 (continued)

2 days of training. Backed out; caught by IDF.

Name: Arin Ahmed
Date: Apr-02
Place: Israel
Age: 20
Status: Student
Notes: From Bethlehem. Volunteered to carry out an attack to avenge the death of her fiance. Supposed to commit bombing during week of April with 16-year-old Issa Badir (who went through with her mission). Ahmed arrested in Jun-02 by IDF.

Name: Shefa'a Alkudsi
Date: Apr-02
Place: Israel
Age: 26
Notes: Divorced with young child. Arrested by IDF on 11-Apr-02.

Name: Unknown
Date: 13-Jun-02
Place: Israel
Age: 15
Organization: Tanzim
Notes: From Bethlehem. Recruited by Tanzim through her uncle.

Name: Unknown
Date: 14-Jun-02
Place: Israel
Notes: Israel Security Forces apprehended 2 female would-be suicide bombers.

Name: Umaya Mohammed Danaj

Date: 27-Jul-02
Place: Israel
Age: 28
Notes: Arrested on her way to commit suicide bombing in Israel.

Name: Zarima Muzhikhoyeva
Date: 11-Jul-03
Place: Russia
Age: 22
Status: Widow
Organization: Chechen
Notes: One lb. of military-issue explosives in bag. Lost husband in fighting against the Russians. Explosives similar to those at rock concert. Policeman killed in attempting to defuse.

Name: Mareta Dudayeva
Place: Russia
Age: 17
Organization: Chechen
Notes: Captured when bomb-laden truck failed to detonate. None of her relatives have died in the war; not religious.

Name: Thawra (Revolution)
Status: Trainee
Notes: Would like to assassinate Ariel Sharon.

Name: Nidal (Struggle)
Status: Trainee

Name: Jihad (Holy War)
Age: 30
Status: Trainee
Notes: Mother of a 12-year-

old son.

Name: Bissam (Smile)
Age: 15
Status: Trainee
Notes: Not to be used as a suicide bomber but being trained in light weapons.

Name: Nour (Light)
Age: 24
Status: Trainee
Notes: Accountant. Saw her sister killed.

Name: Um Sakher (the Mother of Rock)
Status: Trainee

Name: Saber (Patience)
Status: Trainee

Name: Tahereer (Liberation)
Status: Trainee

Name: Hiba
Age: 28
Status: Trainee
Notes: Has five children. Father died in first intifada; husband was killed three years ago while she was pregnant. Two of her brothers have also been killed; brother-in-law is in Iraqi prison.

Name: Leila
Age: 22
Status: Trainee
Notes: Student. Family member lost homes, friends living in tents, 12 brothers and sisters.

Why use suicide bombers?

Terrorism has been defined as "a synthesis of war and theater."[14] This descriptor aptly applies to female suicide bombings. Suicide bombing is used because:

• It is a simple and low-cost operation (requiring no escape route or rescue mission).

• It increases the likelihood of mass casualties and extensive damage (since the bomber can choose the exact time, location, and circumstances of the attack).

• There is no fear that interrogated terrorists will surrender important information (because their deaths are certain).

• It has an immense impact on the public and the media (because it precipitates an overwhelming sense of helplessness).[15]

Others note that suicide terrorism "inflicts profound fear and anxiety and produces a negative psychological effect on an entire population and not just on the victims of the actual attack."[16] Finally, a "suicide attack attracts wide media coverage and is seen as a newsworthy event."[17] Female suicide bombers (youngest, oldest, mother, Red Cross worker, jilted fiancé, avenger of dead family members) have an added media aspect which encourages terrorists to capitalize on the sensationalism.

Organizations which routinely use suicide bombers "have utilized the notion of martyrdom and self-sacrifice as a means of last resort against their conventionally more powerful enemies."[18] These groups believe that suicide bombs are successful in bringing notice to their plight and contend that suicide bombers are the only effective weapons they have, in contrast to their enemies' much larger wealth, weapons, soldiers, and political means. Abu Shanab, a Hamas leader, stated that "all that is required is a bomb, a detonator, and a moment of courage, and courage is the scarce resource."[19]

---

14   Cindy C. Combs and Martin Slann, *Encyclopedia of Terrorism*, New York: Facts on File, Inc., 2002, p. 20.

15   Sprinzak, p. 68.

16   Yoram Schweitzer, "Suicide Terrorism: Development & Characteristics," April 21, 2000. http://www.ict.org.il/articles/articledet.dfm?articleid=112.

17   Ganor, "The First Iraqi Suicide Bombing."

18   Russell D. Howard and Reid L. Sawyer, *Terrorism and Counterterrorism*, Guilford, CT: McGraw-Hill/Dushkin, 2002, p. 130.

19   Stern, *Terror in the Name of God*, p. 40.

Why use female suicide bombers?

The use of women as suicide bombers poses conflicts with some fundamental religious leaders' beliefs, while serving the tactical need for a stealthier weapon. In January 2002, Sheikh Ahmed Yassin, the spiritual leader of Hamas, "categorically renounced the use of women as suicide bombers."[20] In March 2002 after the second Fatah bombing, he reported that "Hamas was far from enthusiastic about the inclusion of women in warfare, for reasons of modesty."[21] That position dramatically shifted on January 14, 2004, when the first Hamas female suicide bomber struck. Why was she used? Yassin defended this change as a "significant evolution in our fight. The male fighters face many obstacles,"[22] so women can more easily reach the targets. He concluded his statement by noting that "Women are like the reserve army – when there is a necessity, we use them."[23]

Terrorist organizations use women as weapons because they provide:

•   Tactical advantage: stealthier attack, element of surprise, hesitancy to search women, female stereotype (e.g., nonviolent).

•   Increased number of combatants.

•   Increased publicity (greater publicity = larger number of recruits).

•   Psychological effect.

"It is the ultimate asymmetric weapon," explained Magnus Ranstorp, director of the Center for the Study of Terrorism and Political Violence. "You can assimilate among the people and then attack with an element of surprise that has an incredible and devastating shock value."[24] In the words of a commander in charge of training future suicide bombers, "The body has become our most potent weapon. When we searched for new ways to resist the security complications facing us, we discovered that our women could be an advantage."[25] One trainer even boastfully described them as the new "Palestinian human

---

20 Arnon Regular, "Mother of Two Becomes First Female Suicide Bomber for Hamas," *Haaretz*, January 16, 2004.

21 Ibid.

22 Ibid.

23 Ibid.

24 Van Natta, sec.4, p. 1.

25 Hala Jaber, "The Avengers," *Sunday Times* (London), December 7, 2003, sec. Features, p. 1.

precision bomb."[26] Those who study Middle East cultures cite another reason: "The use (of female suicide bombers) by Palestinian militant groups is designed to embarrass the Israeli regime and show that things are so desperate that women are fighting instead of men."[27] Suicide bombers provide the low-cost, low-technology, low-risk weapon that maximizes target destruction and instills fear – women are even more effective with their increased accessibility and media shock value.

Who becomes a suicide bomber?

The answer to this question is elusive for both males and females. Some suggest there are definite trends, others dispute that conclusion, and still others maintain that a profile cannot be developed as it is "unlikely that the search would be successful in creating a set of common denominators that could span several continents, time periods, cultures, and political configurations."[28] Some of the factors assessed included age, education, economic status, and socialization toward violence.[29] The only factor that "all the experts seem to agree on is that suicide bombers are primarily young people."[30] Analysts note "The positive attitudes toward political violence – already well entrenched in persons under 17 years of age (14.5 percent) – actually increases in the population up to the age of 24 (14.9 percent) and decreases thereafter (6 percent at 64 years of age)."[31] In addition to age, it appears that education plays a role as "the percentage increases with the level of education: 8.3 percent among those with an elementary school diploma versus 12.8 percent among those with a university degree."[32]

Profiles may be developed for specific sets of suicide bombers. For example, Jessica Stern profiles the typical male Palestinian suicide bomber (young, unmarried, mosque attendee, etc.). Other experts concur, such as

---

26 Ibid.

27 Phillip Smucker, "Arab Women Take to the Streets; Pro-Palestinian Demonstrations in Arab Nations This Week Have Included More Women," *Christian Science Monitor*, April 16, 2002, sec. World, p. 6.

28 Combs and Slann, p. 215.

29 Ibid., pp. 216-218.

30 Ibid., p. 215.

31 Luisella Neuburger and Tiziana Valentini, *Women and Terrorism*, New York: St. Martin's Press, 1996, p. 24.

32 Ibid.

Ariel Merari, a leading Israeli authority on suicide terrorism[33] and Boaz Ganor, Executive Director of the ICT.[34] Additionally, most suicide bombers tend to be of average economic status (although a disproportionate number come from refugee camps).[35] Likewise, "as logical as the poverty-breeds-terrorism argument may seem, studies show that suicide attackers are rarely ignorant or impoverished."[36] Two other critical findings reveal that more than half had spent time in Israeli prisons[37] and "had expressed the desire to avenge the death or injury of a relative or a close friend."[38]

Who becomes a female suicide bomber?

"The reasons for women's participation in deadly attacks vary greatly and it is hard to generalize, for this phenomenon is too recent and the attacks have been too few. Either not enough research has been conducted yet or the sample size is too small to make effective generalizations."[39] Although the data are limited, female suicide bombers, just like male suicide bombers, have one characteristic which typifies all – they are young. The average age varies from 21.5 (Turkey) to 23 (Lebanon), a small differential. Other characteristics do not hold. Some are widows and others have never been married; some are unemployed and others are professionals; some are poor and others are middle class.

Most analysts can easily compare the Black Widows in Russia with the Palestinian suicide bombers, since both appear to be serving "struggles of national identity"[40] with religious overtones. Additionally, as is true of the male counterparts, several female suicide bombers have experienced the loss of a close friend or family member. The selection of women for suicide operations and the methods used to persuade them generally are similar to those employed for men. The recruiters take advantage of the candidates'

---

33 Stern, *Terror in the Name of God*, p. 51.

34 Boaz Ganor, "Suicide Terrorism: An Overview," February 15, 2000, available from http://www.ict.org.il/articles/articledet.cfm?articleid=128.

35 Stern, *Terror in the Name of God*.

36 Scott Atran, "Who Wants to be a Martyr?" *New York Times*, May 5, 2003, sec. A, p. 23.

37 Stern, *Terror in the Name of God*.

38 Ganor, "Suicide Terrorism: An Overview."

39 Beyler.

40 "This Litany of Carnage Explodes the Myth That Terrorism Can Be Easily Defeated," *Independent* (London), May 19, 2003, sec. Comment, p. 14.

innocence, enthusiasm, personal distress, and thirst for revenge.[41]

Finally, the question of when to recruit comes into play. Some believe that recruiters of female suicide bombers try to "get them while they are young as the LTTE leaders personally brainwashed preadolescent boys and girls, many of them orphans."[42] Russian authorities seem to believe that "you need complete control of all their conditioning, usually from the time they are 11 or 12."[43] At the same time, however, the Russians obviate their own argument by noting that the Chechen insurgency hasn't been going on long enough to "grow" the number of female suicide bombers who have attacked them.[44] Other Russians believe that the female suicide bombers are "sold" to terrorist organizations, drugged to perform such acts, and/or raped and blackmailed if they do not participate. Finally, as one journalist describing the Black Widows concluded, "There are so many theories to explain the women's motivation that it is impossible to sort through them."[45]

Why become a suicide bomber?

There are "religious, nationalistic, economic, social, and personal rewards" for suicide bombers.[46] Researchers believe there are "few differences between a man and a woman carrying out such a mission. It may be a surprise, but motivations are the same: they do believe, they are committed, they are patriotic, and this is combined with a religious duty."[47] Religious terrorism is a particularly potent form of violence; religion offers the moral justification for committing seemingly immoral acts. Nationalistic fanatics court suicide bombers and use rhetoric to stir up feelings of patriotism, hatred of the enemy,

---

41 Yoram Schweitzer, "A Fundamental Change in Tactics," *Washington Post*, October 19, 2003, sec. B, p. 3.

42 Jonathan Kay, "Watching Her Go from a Doll to a Rock to a Bomb," *Los Angeles Times*, January 30, 2002, sec. California Metro.

43 Randy Boswell, "Black Widows Put Moscow Under Siege: Female Suicide Bombers Spread Wave of Terror Across Russian Capital," *Ottawa Citizen*, July 11, 2003, sec. A, p. 1.

44 Ibid.

45 Mark McDonald, "Chechnya's Eerie Rebels: Black Widows—the 19 Black-Clad Female Terrorists in Moscow's Theater Siege—Are Still Shrouded in Mystery, One Year Later," *Gazette* (Montreal, Quebec), October 24, 2003, sec. A, p. 4.

46 Ganor, *Suicide Terrorism: An Overview.*

47 Gregg Zoroya, "Her Decision to be a Suicide Bomber," *USA Today*, April 22, 2003, sec. A, p. 1.

and a profound sense of victimization.[48] Suicide bombers then see their own actions as being driven by a higher order; they believe their sacrifice will provide rewards for them in the afterlife.[49] Devout Muslims believe that, in death, every martyr, male or female, is welcomed by a minimum of 70 apparitions (houri-el-ein) of unnatural beauty who wipe away the martyrs' sins, open the gates of heaven, and provide them with all the pleasures that God has given to mankind.[50]

Beyond religious and patriotic motivations, suicide bombers may receive large sums of money, improve their family's social status, and enhance their reputation. After their death, their families are showered with honor and receive substantial financial rewards.[51] Additionally, suicide bombers expect to be admired and envied by those left behind. Photographs capture them in heroic positions, and these photos will be used as recruitment posters.[52] In fact, a "study of world attitudes by the Pew Research Center showed ever-rising support for 'martyrs,' and recruiters tell researchers that volunteers are beating down the door to join."[53]

Finally, many organizations deliberately are targeting women for strategic purposes because female suicide bombers receive more media attention. Research has shown that "public perceptions of the level of terrorism in the world appear to be determined not by the level of violence, but rather by the quality of the incidents, the location, and the degree of media coverage."[54] So the media provides both an advertising and recruitment tool for terrorist groups. Analysts noted that, when the first Palestinian female bombing occurred, the "news was given great prominence... far more than any male suicide bomber would have received. Women who kill or threaten to kill are hot news. It is a reaction that knows no state or religious boundaries."[55] Experts worry that "if this catastrophe leads to more women following suit, this will

---

48 Sprinzak, p. 69.

49 Jeffrey D. Simon, *The Terrorist Trap*, Indianapolis: Indiana University Press, 1994, p. 310.

50 Jaber.

51 Ganor, "Suicide Terrorism: An Overview."

52 Stern, *Terror in the Name of God*.

53 Atran.

54 Yonah Alexander and John M. Gleason, eds., *Behavioral and Quantitative Perspectives on Terrorism*, New York: Pergamon Press, 1981, p. 8.

55 Melanie Reid, "Myth that Women are the Most Deadly Killers of All," *Herald* (Glasgow), January 29, 2002, sec. A, p. 14.

attract disproportionate publicity."[56] This disproportionate publicity, in turn, may arouse worldwide sympathies for suicide bombers and can also serve as a terrorist recruitment tool.

How are female suicide bombers trained?

Until recently, very little has been known about the female suicide bombers' training. Indeed, in February 2003 the ICT published an article which stated that Palestinian female bombers were not trained nor prepared psychologically for suicide attacks.[57] However, the most recent information, which came from women who have either backed out or been unsuccessful in bombing attempts, proved that wrong. Russian Security Forces captured at least two potential female suicide bombers, and the Israeli Army is said to have 17-20 in custody.[58]

Russian officials maintain that the female suicide bombers "were trained abroad by Islamic fundamentalists outside of Russia, are paid for abroad, and are organized by the head of the Arab agents of Al-Qaeda."[59] All potential bombers "are thoroughly trained by Arab psychologists and demolition experts,"[60] and "clandestine training centers have become a conveyor belt in turning out female cannon fodder."[61] Zarima Muzhikhoyeva, the first live shakhidka (martyr), was captured when her purse filled with 1.5 kilograms of military-issue explosives failed to detonate. Ms. Muzhikhoyeva's husband was killed in the first war in Chechnya when she was pregnant with her daughter. A female friend offered to help her and agreed to pay her debts, give her grandparents money, and provide for her daughter – all that was required in return was that she choose the true road to Allah. She was taken by rebel fighters to the mountains for 1 month. There, she cooked and washed the fighters' clothes, prayed daily, and listened to the atrocities of the Russian troops in Chechnya.

When her training was complete, she went to Moscow and was

56 Ibid.

57 Beyler.

58 "The Bomber Next Door," *60 Minutes II*, May 28, 2003.

59 "Female Suicide Bombers Were Trained Abroad—Russian Security Service," BBC Worldwide.

60 Ibid.

61 Argumenty i Fakty, "How Many More Suicide Bombers?" *What the Papers Say* (Russia), July 11, 2003, sec. Shorts, p. 1, 6.

housed with the two young women who later blew themselves up at a rock concert. After her capture, she led police to the house and yard where they found a buried metal container filled with explosive belts.[62] The Russians released little information from her interrogation or from interrogation of another unsuccessful female suicide bomber, Mareta Dudeyeva.

In Israel, 25-year-old Tawriya Hamamra, from the Al Aqsa Martyrs Brigade, discussed the training she received in May 2002 before she decided to abort her mission. Two weeks before her mission, she was sent to Nablus, met by a Fatah official, placed in a student flat, and began training. Four people trained her in two 45-minute sessions on Friday and Saturday. On Saturday she told the trainers that she had changed her mind. Although they were angry, they allowed her to leave. She was arrested the next day in an Israeli raid in the nearby city of Tulkarm.[63]

The most recent and in-depth look at Palestinian training of female suicide bombers was published in December 2003. The reporter met the commander in charge, trainers, and nine future Palestinian female suicide bombers:

> ... Women were handling explosives and familiarizing themselves with Kalashnikov sub-machineguns. Girls are taught to assemble and dismantle their AK-47 assault rifles, and target practice follows – as do hours of theory about the 'enemy and its tactics.'

> The details and outcome of each attack are dissected, revised, debated, and discussed. The women spend as much as 6 hours a day familiarizing themselves with explosives. They are introduced to the bomb belts that will rip their flesh, while killing and maiming those around them. Finally, the girls have to practice moving around with the weight of the explosive belts strapped to their bodies. Sometimes the explosives are distributed around the body; some strapped to their legs, others to their backs or abdomens. "It all depends on the build and shape of the woman and how best to strap her without over-bulging parts of her body," a male handler explained.[64]

Implications of change

In March 2002, after an 18-year-old female killed two innocent people in a supermarket in Israel, a professor of Middle East Studies commented:

62   Luba Vinogradova, "Deadly Secret of the Black Widows," *Times* (London), October 22, 2003, sec. Features, p. 4.

63   Matthew Kalman, "Inside the Mind of a Suicide Bomber; Recruit Reveals Training of Females for Mission," *Standard* (St. Catharines), May 31, 2002, sec. C, p. 10.

64   Jaber.

"I have a great fear that, if this continues, they will pull in younger and younger women. Once they break the boundaries of what is accepted on a human level, there are no boundaries."[65] For female suicide bombers, there appear to be few boundaries left to cross.

From a strategic perspective, the most significant change has been the clergy's approval of female martyrs. Suicide bombers believe they are martyrs and are told they are the "true defenders of the oppressed and dispossessed."[66] Many experts perceive the problem to be the "blurring of differences and the increased confusion between nationalism and religiosity."[67] Until the Chechen terrorist campaign, "the use of female suicide bombers was a clear indicator of secular terrorism. The growth in the number of Chechen female suicide bombers signaled the beginning of a change in the position of fundamentalist Islamic organizations regarding the involvement of women in suicide attacks."[68]

Some fundamentalist clerics have expressed theological dilemmas in sanctioning suicide bombing,[69] as Islam explicitly forbids suicide (*inithar*). The Koran states: "And do not kill yourself, for God is indeed merciful to you."[70] Others note that the Prophet Mohammed explicitly forbade suicide and decreed, "He who drinks poison and kills himself will carry his poison in his hand and drink it in Hell for ever and ever."[71] Militant religious leaders thus have asserted that these attacks are acts of martyrdom (istishhad) as a means of last resort.[72] And while God punishes those who commit suicide, he rewards the martyr. According to the Koran, "Think not of those who are slain in the cause of God as dead. Nay, they are alive in the presence of the Lord and are granted gifts from him."[73]

"According to the level of religiosity, terrorist organizations have different policies concerning women suicide bombers. Saudia Arabia originally

---

65 Margo Harakas, "Palestinian Women Defy Islamic Tradition to Become Suicide Bombers," *Sun-Sentinel*, April 15, 2002, sec. Lifestyle, p. 10.

66 Howard and Sawyer, p. 129.

67 "This Litany of Carnage."

68 Schweitzer.

69 Howard and Sawyer, p. 130.

70 Stern, *Terror in the Name of God*, p. 52.

71 Jonathan Kay, "Defaming Islam—One Bomb at a Time," *National Post* (Canada), July 16, 2003, sec. A, p. 14.

72 Howard and Sawyer.

73 Stern, *Terror in the Name of God*.

refused to legitimize female suicide bombings as martyrdom; however, in August 2001, the High Islamic Council in Saudi Arabia issued a fatwa encouraging Palestinian women to become suicide bombers."[74] Lebanese Muslim cleric Sheikh Mohammed Hussein Fadlallah declared, "It is true that Islam has not asked women to carry out jihad (holy war), but it permits them to take part if the necessities dictate that women should carry out regular military operations or suicide operations."[75] The Koran states that jihad can be carried out by women as well as men, but "most contend that women can serve as combatants only after the male ranks have been depleted. This is the demarcation that had kept Hamas from recruiting women."[76] In January 2004, that demarcation disappeared.

Another strategic concern is the possibility that female suicide bombing is being supported, financed, and directed by a global terror network. Some believe that the "resort to martyrdom can be explained by an increasing level of internationalization between groups in terms of contact, similarity of causes, and examples of strategies."[77] In the Chechen terrorist campaign, "while the groups involved act in the name of an ethnic-nationalist ideology, they are thought to be cooperating with global jihad organizations connected to Al-Qaeda and count several 'Afghan alumni' among their commanders."[78]

Additionally, in March 2003, "the Saudi-owned *Asharq Al-Awsat* published an e-mail interview with a leader of the female mujahedeen of Al-Qaeda. She told them her instructions came from Al-Qaeda and the Taliban, mainly via internet."[79] The woman went on to state that the organization was planning "a new attack which would make the United States forget September 11, and that the idea came from the martyr operations carried out by the Palestinian women." She further outlined training methods and potential locations. This communiqué is especially interesting because currently Al-Qaeda has "... refrained from including women in their operations except in supporting roles. The religious and ideological leap required [to use female suicide bombers] apparently represents a daunting challenge for Al-Qaeda. But if the outcome

---

74 Beyler.

75 "Lebanese Muslim Cleric OK's Female Suicide Bombers," *Business Recorder*, April 2, 2002.

76 Harakas.

77 Howard and Sawyer, p. 128.

78 Schweitzer.

79 "Bin Laden Sets up Female Suicide Squads to Bomb U.S.," Agence France Presse, March 12, 2003, sec. International News.

seems worthwhile, there is a real prospect that Al-Qaeda, too, will cross the Rubicon, and then find some religious justification for it."[80]

Strategic assessment

"Although profiling suicide bombers may be a fascinating academic challenge, it is less relevant in the real-world struggle against terrorism than understanding the modus operandi of terrorist leaders."[81] Some analysts are concerned most with structural issues: "The most important factor is the organization: almost nobody does this as an individual."[82] They point out the need to counter terrorist organizations: "Since suicide terrorism is an organizational phenomenon, the struggle against it cannot be conducted on an individual level."[83] Most important, "organizations only implement suicide terrorism if their community approves of its use."[84] These organizations, however, are not equally reliant on suicide bombers: "A careful study of all the organizations that have resorted to terrorism since 1983 suggested that the most meaningful distinction among them involves the degree to which suicide bombing is institutionalized."[85] Brian Jenkins, a terrorism expert, agrees. "I think we'd all agree that suicide bombing is abnormal. The fact that abnormal behavior is applauded in the community reflects abnormal conditions. If normal conditions are restored, then normal behavior should return – at least they would be less tolerant of abnormal behavior."[86]

But why would a community view suicide bombing as normal behavior? Some blame military occupation and/or deplorable economic and social conditions.[87] While those may be valid considerations, why haven't all cultures that have endured similar experiences supported suicide bombers? Brian Barber, a psychologist, interviewed 900 young adults from Gaza and a comparison group of Bosnian Muslims "who had also suffered through violence but had not become a source of suicide bombers. Faith was the largest difference:

---

80   Schweitzer.

81   Sprinzak, p. 69.

82   Stern, *Terror in the Name of God*, p. 51.

83   Sprinzak.

84   Ibid., p. 72.

85   Ibid., p. 69.

86   Van Natta.

87   Combs and Slann, p. 84.

the Palestinians routinely invoked religion to invest personal trauma with social meaning, whereas Bosnians did not consider religion significant to their
lives."[88] This is a critical distinction: "The strictly hierarchical nature of religious terrorist groups with a highly disciplined structure and obedient cadre
means not only that the main clerical leaders command full control over the
political as well as military activities of the organization, but also the strategies
of terrorism are unleashed in accordance with general political directives and
agendas."[89]

        Even if the community supports this weapon which is legitimized by
the religious hierarchy, why are suicide bombings increasing, especially among
the young? One explanation is that suicide bombers are seen as successful because their service is for a higher cause;[90] this success "incentivizes those who
recruit and send the suicide bombers on their lethal missions."[91] Jessica Stern
proposes the answer might be "social contagion:" Ordinary suicide has been
shown to spread through social contagion especially among youth. Studies
have shown that a teenager whose friend or relative commits suicide is more
likely to commit suicide himself. Suicide bombing entails a willingness not
only to die, but also to kill others. The situation in Gaza suggests that suicide-
murder can also be spread through social contagion. "Martyrdom operations"
are part of the popular culture. For example, on the streets of Gaza, children
play a game call shuhada, which includes a mock funeral for a suicide bomber.
Teenage rock groups praise martyrs in their songs. Asked to name their he-
roes, young Palestinians are likely to include suicide bombers.[92]

        Research indicates that terrorist organizations will continue to use
suicide bomber tactics and employ female suicide bombers. A comprehensive
counterterrorism plan should recognize the increasing potential for use of suicide bombers, including females. Specifically, the United States must continue
to lead the way in this fight with terrorism. The United States has the capabilities (diplomatic, economic, information, military) to provide leadership in
combating these problems.

        Diplomatically, the United States should continue to obtain interna-

---

88   Atran.

89   Howard and Sawyer, p. 128.

90   Combs and Slann, p. 213.

91   Alan M. Dershowitz, *Why Terrorism Works*, New Haven, CT: Yale University Press, 2002, p.
26.

92   Stern, *Terror in the Name of God*, pp. 52-53.

tional and United Nations (UN) support in public designation of terrorists, as well as improved information sharing, tightened border security, and reduction of terrorist financing. The UN Security Council adopted Resolution 1373, which requires all states to prevent and suppress the financing of terrorist acts, to include freezing funds and financial assets. In the words of U.S. Secretary of State Colin Powell, "Diplomacy helps us to take the war to the terrorist, to cut off resources they need and depend upon to survive."[93] Reducing their funding and tightening worldwide controls should reduce the global reach of terrorist organizations that employ suicide bombers. Additionally, since the most recent bombings have occurred as a result of the Israeli-Palestinian conflict, the United States needs to push for peaceful resolution of problems in that area.

Worldwide terrorism uses people as bombs under the guise of addressing human problems, so "prevention will require a great improvement in the social conditions that produce, beget, or trigger terrorist acts."[94] Economically, the United States should help develop an "effective counterterrorist strategy that would seek to reduce the terrorist group's base of support through development programs."[95] This aid would provide economic, material, and psychological support to those who have identified with terrorist groups. One major reason Hamas has been so successful is because it provides "social aid" that assists the poor, provides medical care, and educates the children. However, it has also been proven that many of these organizations are merely fronts for funneling money to terrorist organizations. "Fundraising, training, indoctrination, and criminal activities are usually covered by all kinds of harmless organizations ranging from religious to humanitarian institutions."[96] These fronts must be identified, and their funding streams eliminated. At the same time, the true humanitarian need can be met by recognized and authentic relief agencies.

On the information front, several actions are appropriate. First, as noted above, suicide bombers are part of a larger organization. These orga-

---

93 Congress, House, International Relations Committee, Hearing of the International Terrorism, Nonproliferation, and Human Rights Subcommittee of the House International Relations Committee, March 26, 2003.

94 Yonah Alexander and John M. Gleason, eds., *Behavioral and Quantitative Perspectives on Terrorism*, New York: Pergamon Press, 1981, p. 30.

95 Rex A. Hudson, "The Sociology and Psychology of Terrorism: Who Becomes a Terrorist and Why?" Library of Congress, September, 1999.

96 Eric Herren, "Counter-Terrorism Dilemmas," April 15, 2002, http://222.ict.org.il/articles/articledet.cfm?articleid=432.

nizations and their support of suicide terrorists should be made known and roundly castigated by the international community. Their religious rationale should be debunked. They should be condemned widely for attacking non-combatants and for the overwhelming viciousness of their attacks. Suicide bombers should no longer be seen as "victims" but as "assassins." The key is to change peoples' perceptions and thus to deny support for them.

Changes in behavior will occur when the community no longer views suicide bombings as accepted, normal – even heroic – behavior. The United States and humanitarian organizations should review the education of younger children in schools that are supported by terrorist organizations, since "pictures of training camps showing small children learning to become suicide bombers have had great impact on public opinion in the international audience."[97] Further, "Goals of a long-range policy should include deterring alienated youth from joining a terrorist organization in the first place. A counterstrategy could be approached within a framework of advertising and civic-action programs which would help to discredit the terrorist group and have a negative impact on their recruitment efforts."[98] For "Without popular support, a terrorist group cannot survive."[99]

The United States should not respond to suicide bombers militarily unless it has specific, credible information that directly links a given organization to terrorists and suicide bombers; however, given that suicide bombers want to die, attacking them militarily may only encourage others to take up the cause. Finally, Israel repeatedly has struck against the Palestinians—destroying suicide bombers' homes, displacing their families, militarily retaliating. To date, Israel military responses have engendered an escalating, ever-intensifying hostile reaction. Most importantly, "the question of the moral legitimacy of counterterrorist activities must be analyzed accordingly. How far can a state go in combating terrorism without risking endangering its democratic structure?"[100]

Operationally, the United States must improve its intelligence capability, using both human and technical information gathering. "Intelligence is the first link in the chain of thwarting any terror attack, but is of the utmost

97   Ibid.

98   Hudson.

99   Ibid.

100   Herren.

importance in thwarting suicide attacks before they are put into practice."[101] In dealing with suicide attacks, the United States should rely on "more and better human intelligence enabling us to penetrate the movement's armies, monitor its recruitment drives, and predict its evolution—including the type of personnel it will recruit."[102] Our counterterrorism strategies should not "rely on standard operation procedures such as race and gender-based profiling,"[103] because doing so "puts the safety of the American people at risk."[104]

Given the globalization of terror from nonstate actors, experts believe that suicide bombings, which will include female suicide bombers, will increase. The next "first" will be the "first" Al-Qaeda female suicide bomber. "The Federal Bureau of Investigation has acknowledged that Al-Qaeda is actively recruiting women,"[105] perhaps for just this role. International terrorist organizations will assess U.S. security vulnerabilities and exploit weaknesses. While this terrorist weapon may not change the political outcome for the United States or its allies, it will negatively and emotionally impact American lives, beliefs, and morale.

---

101 Ganor, "Suicide Terrorism: An Overview."

102 Stern, "When Bombers are Women."

103 Ibid.

104 Ibid.

105 Ibid.

14

The Decisive Weapon: The Brigade Combat Team Commander's Perspective on Information Operations

BG Ralph O. Baker, USA

This article was solicited from the author by the editor in chief of Military Review subsequent to a briefing the author presented to the Information Operations Symposium II held at Fort Leavenworth, Kansas, on 15 December 2005. The text is an edited version of a transcript from that briefing. It includes additional material and clarification of facts and events provided by the author.

Duty in Iraq has a way of debunking myths and countering "Ivory Tower" theories with hard facts on the ground. I admit that while I was preparing to serve in Iraq as a brigade commander, I was among the skeptics who doubted the value of integrating information operations (IO) into my concept of operations. Most of the officers on my combat team shared my doubts about the relative importance of information operations. Of course, in current Army literature there is a great deal of discussion about IO theory. There is significantly less practical information, however, that details how theory can be effectively translated into practice by tactical units. My purpose in writing this article is to provide commanders the insights I gleaned from my experience.

Soon after taking command of my brigade, I quickly discovered that IO was going to be one of the two most vital tools (along with human intelligence) I would need to be successful in a counterinsurgency (COIN) campaign. COIN operations meant competing daily to influence favorably the perceptions of the Iraqi population in our area of operations (AO). I quickly concluded that, without IO, I could not hope to shape and set conditions for my battalions or my soldiers to be successful.

It certainly did not take long to discover that the traditional tools in my military kit bag were insufficient to compete successfully in this new operational environment. As a brigade commander, I was somewhat surprised to find myself spending 70 percent of my time working and managing my intelligence and IO systems and a relatively small amount of my time directly involved with the traditional maneuver and fire support activities. This was a paradigm shift for me. The reality I confronted was far different from what I had professionally prepared for over a lifetime of conventional training and experience.

Background

My brigade, the 2nd Brigade Combat Team (BCT), was part of the 1st Armored Division. For the first twelve months in Iraq, we were task organized in Baghdad with up to eight battalions, roughly 5,000 strong, all trained for conventional combat. The BCT consisted of two mechanized infantry battalions, a cavalry squadron, an armor battalion, a field artillery battalion, an engineer battalion, a support battalion, and a military police battalion. At headquarters were staff enablers such as psychological operations (PSYOP) and civil affairs (CA) detachments. At one point, my task organization also included 12 U.S. Army National Guard or Reserve Component companies.

My brigade's AO covered roughly 400 square kilometers and encompassed two of the nine major districts in Baghdad: Karkh and Karada. In those two heavily populated and congested districts lived between 700,000 to a million citizens. The area contained at least 72 mosques and churches. In the northwest part of our AO, the population was predominantly Sunni. This area also contained a small neighborhood called Kaddamiya, where Saddam Hussein had grown up. Not surprisingly, that community was a bastion of staunchly pro-Baath sentiment and was steadfastly loyal to Saddam. Such demographic factors made that part of our AO particularly volatile and problematic.

In contrast, our area also contained the Karada district, one of the most affluent parts of the city. Three universities are located there, Baghdad University being at the very southeastern tip. Many Western-trained and educated elites live in Karada, and many of Baghdad's banks and headquarters for major businesses are there. The population in this area is characteristically more secular in its views and somewhat more receptive to outside ideas and influence. In addition, 70 percent of the embassies and diplomatic residences in Baghdad were situated in our AO (figure 1).

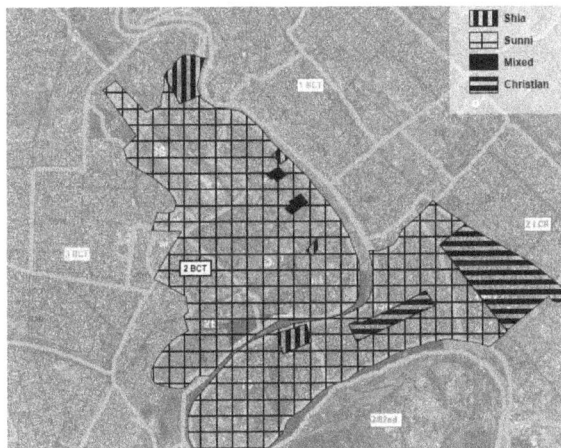

Figure 1.
2BCT/1AD
battlespace religious
demographics:
Karkh.

The southeastern region of our area was home to a principally Shi'ite population. The infrastructure in this area was, in comparison to other parts of the city, shabby. In many places the population lived in almost uninhabitable conditions, the neighborhoods having been largely neglected by the Baathist regime for years (figure 2). Another significant component of this complex society was the Christian population. Baghdad has the largest Christian population in the country, and it was also concentrated inside our battlespace.

The demographic diversity in 2nd Brigade's AO produced a lot of different ethnic, cultural, and religious dynamics. Consequently, each area presented unique IO challenges. And, of course, this already complex situation was made more complex by insurgent and terrorist violence and the persistent lack of infrastructure and basic services.

Also of note was what proved to be an additional geographic area

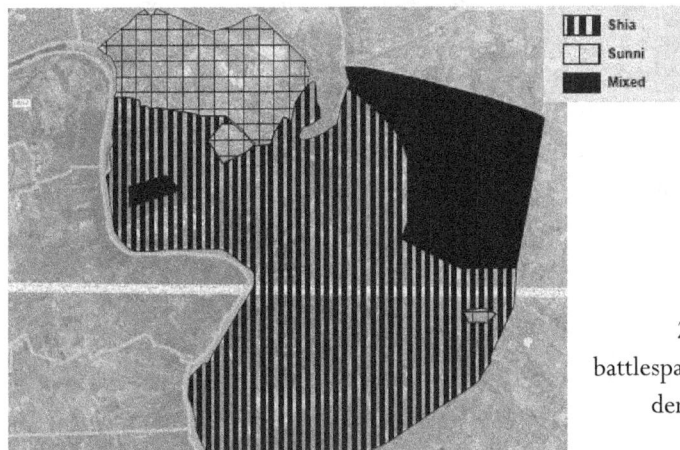

Figure 2.
2BCT/1AD
battlespace religious
demographics:
Karada.

with a completely different IO population of interest, one that had its own set of parochial concerns and priorities: the Green Zone. This area housed the headquarters of the Coalition Provisional Authority and Combined Joint Task Force 7.

Another vital demographic, one that my commanders and I found we had inadvertently taken for granted and failed to address effectively, was our own soldiers. Most news that soldiers typically received came from watching CNN, the BBC, or Fox News. Soldiers were getting the same inaccurate, slanted news that the American public gets. With a significant amount of negative news being broadcast into their living quarters on a daily basis, it was difficult for soldiers to realize they were having a positive impact on our area of operations.

Once we appreciated the dynamics of the demographics in our AO, we found that we could easily fit Iraqi citizens into three broad categories: those who would never accept the Coalition's presence in Iraq (religious fundamentalists, insurgents, terrorists); those who readily accepted the Coalition's presence in Iraq (typically secular, Western-educated pragmatists); and the vast majority of Iraqis, who were undecided. We referred to this last category as the silent majority and focused much of our information operations on influencing this group.

Adjusting the plan to IO realities

One of the first challenges I faced was to understand the overarching IO plan for Iraq and, more importantly, how my combat team was supposed to support it. Part of the challenge at this time for everyone – battalion through corps – was our lack of IO experience and our ignorance of how valuable IO is to COIN success. In fact, during the summer of 2003 there was still much debate over whether or not we were even fighting an insurgency. The IO support we did receive from higher headquarters included broad themes and messages that we were directed to communicate to the local populations. Unfortunately, these messages were often too broad to resonate with the diverse subpopulations within brigade and battalion areas.

This brings me to my first essential IO observation: To be effective, you must tailor themes and messages to specific audiences. IO planners at commands above division level appeared to look at the Iraqis as a single, homogeneous population that would be receptive to centrally developed, all-purpose, general themes and messages directed at Iraqis as a group. In many cases, the guidance and products we received were clearly developed for a high-level diplomatic audience and were inappropriate or ineffective for the diverse populations clustered within our battalion AO.

When we did request and receive theme support or IO products, they were typically approved too late to address the issue for which we had requested them. To overcome what was an ineffective and usually counter-productive attempt by the IO/PSYOP agencies at higher levels of command to centrally control themes and messaging, we were compelled to initiate a more tailored IO process. We developed products that incorporated relevant themes and messages fashioned specifically for the diverse groups and microp-opulations in our area of operations.

A guiding imperative was to produce and distribute IO products with focused messages and themes more quickly than our adversaries. Only then could we stay ahead of the extremely adroit and effective information opera-tions the enemy waged at neighborhood and district levels.

We were also initially challenged in working through the bureaucratic IO/PSYOP culture. We often faced situations where we needed handbills spe-cifically tailored to the unique circumstances and demographics of the neigh-borhoods we were attempting to influence. However, the PSYOP community routinely insisted that handbills had to be approved through PSYOP channels at the highest command levels before they could be cleared for distribution. This procedure proved to be much too slow and cumbersome to support our IO needs at the tactical level.

Good reasons exist for some central control over IO themes and products under some circumstances, but information operations are Operations, and in my opinion that means commander's business. IO is critical to combating an insurgency successfully. It fights with words, symbols, and ideas, and it operates under the same dynamics as all combat operations. An old Army saw says that the person who gets to the battle the "firstest" with the "mostest" usually wins, and this applies indisputably to information operations. In contrast, a consistent shortcoming I experienced was that the enemy, at least initially, consistently dominated the IO environment faster and more thoroughly than we did. Our adversary therefore had considerable success in shaping and influencing the perceptions of the Iraqi public in his favor. The ponderous way in which centrally managed PSYOP products were developed, vetted, and approved through bureaucratic channels meant they were simply not being produced quickly enough to do any good. Just as importantly, they were not being tailored precisely enough to influence our diverse audiences' opinions about breaking events.

Faced with bureaucratic friction and cumbersome policy, and thrust into an IO arena quite different from that for which most of us had been trained, I had to make decisions concerning IO matters based on common sense and mission requirements. To this end, I consciously had to interpret policy and regulatory guidance in creative ways to accomplish the mission as we saw it, though in a manner such that those who wrote the original regulations and guidance probably had not intended. This was necessary because Cold War regulations and policies were holding us hostage to old ideas and old ways of doing business. They were simply no longer valid or relevant to the challenges we were facing in this extremely fluid, nonlinear, media-centric COIN environment that was Baghdad circa 2003-2004.

Of course, such an approach made some people uncomfortable. As a rule, if our application of IO techniques was perceived to violate a strict interpretation of policy or regulation, I asked myself: Is it necessary to accomplish our mission, and is our tactic, technique, or procedure morally and ethically sound? If the answer was yes, I generally authorized the activity and informed my higher headquarters.

We were not a renegade operation, however. If what we thought we had to do ran counter to written policies and guidance, I kept my division commander informed in detail of what, when, and why we were doing it. Fortunately, the command environment was such that initiative, innovation, and common-sense pragmatism were supported in the face of uncertainty and lack of relevant doctrine. One example of this sort of support was our decision to adopt, as a policy, the engagement of foreign, Iraqi, and international media at the earliest opportunity following a sensational act of insurgent violence.

The guidance within which we were operating was that brigades could not conduct press conferences. In my view, that policy was counterproductive. Headquarters above division were usually slow to react to major events involving terrorism on the streets, and costly hours would go by without an appropriate public response to major terrorist incidents. We experienced firsthand the detrimental effects that this ceding of the information initiative to insurgents was having in our area. The Iraqis had increasingly easy access to TV and radio, but restrictions prevented us from engaging those media to communicate our public information messages rapidly, efficiently, and directly at critical times. By contrast, press reports appeared quickly in the Arab media showing death and destruction in great detail, which undermined confidence in the ability of the Iraqi Provisional Council and the Coalition to provide security.

Our adversary also frequently twisted media accounts in a way that successfully assigned public blame to the Coalition – and the 2nd Brigade specifically – for perpetrating the violent attacks. When slow IO responses and outright public information inaction in the face of such incidents dangerously stoked public discontent, we decided to engage the media on our own in order to get the truth out to the multitudes of people living in our area. If we were going to influence our silent majority successfully, we were going to have to convince them that it was in their best personal and national interest to support the Coalition's efforts. We had to convince them that the insurgents and terrorists were responsible for harming Iraqi citizens and inhibiting local and national progress.

As an illustration, on 18 January 2004 a suicide bomber detonated a vehicle-borne improvised explosive device (VBIED) during morning rush hour at a well-known Baghdad checkpoint called Assassin's Gate, a main entrance into the Green Zone. This attack killed about 50 Iraqis waiting at the checkpoint. While we were managing the consequences of the incident, which included dealing with a considerable number of international and Arab media, I was instructed not to release a statement to the press – higher headquarters would collect the facts and release them at a Coalition-sponsored press conference to be held at 1600 Baghdad time.

Unfortunately, the terrorists responsible for this bombing were not constrained from engaging the press. While precious time was being spent "gathering facts," the enemy was busily exploiting to their advantage the ensuing chaos. The message they passed to the press was that Coalition soldiers were responsible for the casualties at the checkpoint because of an overreaction to somebody shooting at them from the intersection; that is, the terrorists were spreading a rumor that the carnage on the street was not the result of a VBIED but, rather, the result of an undisciplined and excessive use of force by my Soldiers.

As precious time slipped by, and with accusations multiplying in the Arab media and tempers heating up, we made a conscious decision that our field grade officers would talk to the press at the site and give them the known facts; in effect, we would hold a stand-up, impromptu press conference. We also decided that in all future terrorist attacks, the field grade officers' principal job would be to engage the press – especially the Arab press – as quickly as possible while company grade officers managed the tactical situation at the incident site.

Subsequently, when such incidents occurred, we took the information fight to the enemy by giving the free press the facts as we understood them as quickly as we could in order to stay ahead of the disinformation and rumor campaign the enemy was sure to wage. We aggressively followed up our actions by updating the reporters as soon as more information became available. As a result, the principal role of field grade officers at incident sites was to engage the press, give them releasable facts, answer questions as quickly and honestly as possible with accurate information, and keep them updated as more information became known.

Our proactive and transparent approach proved to be an essential tool for informing and influencing the key Iraqi audiences in our AO; it mitigated adverse domestic reaction. Our quick response helped dispel the harmful rumors that nearly always flowed in the wake of major incidents. I heard that the methods we were using with the media immediately following such incidents caused considerable hand-wringing and resentment in some circles. However, no one ever ordered us to stop, no doubt because the positive effects were clearly apparent.

Executing our IO plan

My second IO observation is that you have no influence with the press if you do not talk to them. Moreover, trying to ignore the media by denying them access or refusing to talk can result in the press reporting news that is inaccurate, biased, and frankly counterproductive to the mission. Not talking to the press is the equivalent of ceding the initiative to the insurgents, who are quite adept at spinning information in adverse ways to further their objectives.

The way we adapted to working with the media contrasted significantly with our initial approach. At first, we allowed reporters to come into our unit areas and, essentially, wander around. What resulted was hit or miss as to whether reporters would find a good theme to report on or whether they would stumble onto something they did not understand and publish a story that was out of context or unhelpful. When this happened, we would scratch our heads and say, "Gee, these press guys just don't get it." Actually, we were the ones not getting it. We lacked a good plan on how to work with the press and interest them in the really great things happening in our area.

Recognizing this, we set about preparing our spokespersons and soldiers to engage the media in a systematic, deliberate manner. We became familiar with what the media needed to know and adept at providing the information they required as quickly as possible. At the same time, we ensured that the messages and supporting themes we felt were important were getting out.

To impress on our leaders and Soldiers the need for a press-engagement strategy, we emphasized agenda-setting. I conveyed the manner in which I wanted my leaders to approach this issue by asking how many of them would just let me go down to their motor pools and walk around without them grabbing me and at least trying to get me to look at the positive things they wanted to show me (while also trying to steer me away from the things that were perhaps "still a work in progress"). I told them: "All of you guys understand and do that. So from now on, when working with the media, adopt this same kind of approach."

Meeting Iraqi expectations

One of the more difficult credibility challenges we encountered among the Iraqis was a consequence of the initial mismanagement of Iraqi expectations before we ever crossed the berm into Iraq. As a result, we were met with enormously unrealistic expectations that we had to manage and were simply unable to gratify in a timely manner. Such expectations grew out of Coalition pronouncements before soldiers arrived that extolled how much better off the average Iraqi citizen's life was going to be when Saddam and his regime were gone.

The concept of "better" proved to be a terrible cultural misperception on our part because we, the liberators, equated better with not being ruled by a brutal dictator. In contrast, a better life for Iraqis implied consistent, reliable electricity; food; medical care; jobs; and safety from criminals and political thugs. When those same Iraqis were sitting in Baghdad in August 2003 suffering 115-degree heat with no electricity, an unreliable sewage system, contaminated water, no prospects for a job, lack of police security, periodic social and economic disruption because of insurgent attacks, and no income or pensions with which to support their families, better had become a problematic concept. It took on the psychic dimensions of having been betrayed by the Coalition.

Unfortunately, this view was exacerbated by the average Iraqi's man-on-the-moon analogy: If you Americans are capable of putting a man on the moon, why can't you get the electricity to come on? If you are not turning the electricity on, it must be because you don't want to and are punishing us.

We came to realize that any chance of success with information operations was specifically tied to immediate, visible actions to improve the average Iraqi's quality of life. Until there was tangible improvement that the Iraqis could experience and benefit from firsthand, lofty pronouncements about how much better life would be under democratic pluralism, as well as the value of secular principles of tolerance and national unity, were meaningless. This leads to my third IO observation: There is a direct correlation between our credibility and our ability to improve demonstrably the quality of life, physical security, and stability in a society. Until we could do the latter, we would continue to lack credibility. This was especially true because we were agents of change from a Western world the Iraqis had been taught to hate virtually from birth.

Reaching out to the community

Iraqis in general had little visibility of the positive aspects of the Coalition and U.S. presence in the country. Positive economic, political, and social reforms and improvements in the security environment generally went unnoticed. Collectively, the Iraqis were simply getting too little information on the good things being accomplished. International and Arab media failed to report favorable news, and little information was being passed by word of mouth. Meanwhile, efforts by Coalition forces to share information were limited because we lacked credibility and because many Iraqi citizens did not understand the horrific toll the insurgency was exacting on Iraqi lives and how much it was affecting infrastructure repair. The problem was that we did not have a coordinated, deliberate plan at the brigade level to provide timely, accurate, focused information to communicate these facts. This changed as we developed an IO concept based on a limited number of themes supported by accurate, detailed messages delivered repetitively to key target audiences.

Preventing IO fratricide

Our brigade IO effort did not begin as a centrally coordinated program within my BCT but, rather, evolved as our understanding of the importance of synchronized IO activities matured. Initially, well-intentioned commanders, many of whom lacked clearly defined brigade guidance, had independently arrived at the same conclusion: They needed an IO plan. Each had therefore begun developing and executing his own IO effort. On the surface this was fine: great commanders were using initiative to solve problems and accomplish the mission. Unfortunately, because our activities were not coordinated and synchronized, we often disseminated contradictory information.

For example, one battalion IO message might state that a recent operation had resulted in the capture of ten insurgents with no civilian casualties. Referring to the same operation, an adjacent battalion might inform its Iraqi citizens that five insurgents had been captured and three civilians accidentally injured. From the Iraqi perspective, because our information was inconsistent, we were not being honest.

One of our major objectives was to earn the Iraqis' trust and confidence. If we continued to contradict ourselves or provide inaccurate information, we would never achieve this goal. We termed this phenomenon of contradictory IO statements "IO fratricide." The remedy for this challenge leads to a fourth significant IO observation: a major IO goal at tactical and operational levels is getting the citizens in your AO to have trust and confidence in you.

We have all heard about "winning hearts and minds." I do not like this phrase, and I liked it less and less as experience taught me its impracticality. The reality is that it will be a long, long time before we can truly win the hearts and minds of Arabs in the Middle East. Most of the people have been taught from birth to distrust and hate us. Consequently, I did not like my soldiers using the phrase because it gave them the idea that to be successful they had to win the Iraqis' hearts and minds, which translated into attempts at developing legitimate friendships with the Iraqis. However, in my view, even with considerable effort it is possible to cultivate friendships with only a small segment of the Iraqis with whom we have frequent contact.

Unfortunately, befriending a small portion of the population will not help us convince the remaining Iraqi citizens to begin tolerating or working with us. For us, given the amount of time we had to influence our target population, the more effective plan was to prioritize our efforts toward earning the grudging respect of our target population within the twelve months we would occupy our AO. This was a more realistic goal. If we could demonstrate to our population that we were truthful and that we followed through on everything we said we would, then we could earn the respect of a population and culture that was predisposed to distrust us.

Conversely, I felt that it would take considerable effort and time (resources we did not have) to develop legitimate friendships – assuming friendships were possible on a broad scale. So, by replacing "winning the hearts and minds of the Iraqis" with "earning the trust and confidence of the Iraqis," I attempted to provide a mental construct to guide our Soldiers and leaders in all aspects of the IO campaign. Subsequently, we began to formulate a general concept for IO based on the objective of garnering the trust, confidence, and respect, however grudging, of the various populations. Our overarching goal was to convince the silent majority that their personal and national interests resided with the Coalition's efforts, not with the insurgents'. If we were to succeed, it was imperative to drive a wedge between the insurgents and the Iraqi population.

Manning the IO cell

Staffing an IO cell at brigade level was another challenge. Because we were not authorized many of the military occupational specialties necessary to plan, coordinate, and control information operations, we built our own IO working group (IOWG) out of hide. Our IOWG consisted of senior officers from the PSYOPs and CA detachments attached to the brigade, one intelligence officer detailed to serve as our public affairs officer (PAO), an engineer officer, and the brigade fire support officer.

The engineer officer was key because much of the visible progress we were enjoying in our AO was the result of renovation and reconstruction activities. The engineer officer maintained visibility on these projects to ensure that we did not miss opportunities to inform the Iraqis of any progress.

Adding a PAO to the IOWG was an obvious step. Because of the immense interest in our operations shown by international and Arab media, I had to assign this duty full time to one of my most competent and articulate officers. Subsequently, we realized that we needed to expand our public affairs activities and therefore hired two Iraqi citizens with media experience to manage our activities with the Arab press.

In concert, we leveraged the doctrinal knowledge of our PSYOPs and CA officers to organize activities and develop messages and distribution concepts. Finally, because our IO activities were ultimately "targeting" specific demographic elements in our AO, it was a natural fit to place the brigade fire support officer in charge of the IOWG.

Evolving unity of effort

Our approach to conducting IO evolved over time, out of the

operational necessity to accomplish our mission. We were probably a good three to four months into our tour before we gained the requisite experience and understanding of key IO factors. We then began to develop deliberately a structure and mechanism to synchronize our information operations systematically throughout the brigade. The following observations ultimately helped shape our operational construct:

• It is imperative to earn the trust and confidence of the indigenous population in your AO. They might never "like" you, but I am convinced you can earn their respect.

• To defeat the insurgency, you must convince the (silent) majority of the population that it is in their best personal and national interest to support Coalition efforts and, conversely, convince them not to support the insurgents.

• For information operations to be effective, you must have focused themes that you disseminate repetitively to your target audience.

• Target audiences are key. You should assume that the silent majority will discount most of the information Coalition forces disseminate simply because they are suspicious of us culturally. Therefore, you must identify and target respected community members with IO themes. If you can create conditions where Arabs are communicating your themes to Arabs, you can be quite effective.

• Being honest in the execution of information operations is highly important. This goes back to developing trust and confidence, especially with target audiences. If you lose your credibility, you cannot conduct effective IO. Therefore, you should never try to implement any sort of IO "deception" operations.

Commander's vision and guidance

Visualizing and describing a concept of operation, one of a commander's greatest contributions to his organization, was a contribution I had yet to provide to my combat team. It was essential to do so immediately. I also understood that after developing an IO plan, I would have to act energetically to ensure that subordinate commanders embraced information operations and executed them according to my expectations. I did, and they embraced the concept and ultimately improved on it. My fifth IO observation is that for all types of military operations, the commander's vision and intent are essential, but when directing subordinate commanders to perform outside of their

Figure 3. 2BCT IO campaign plan.

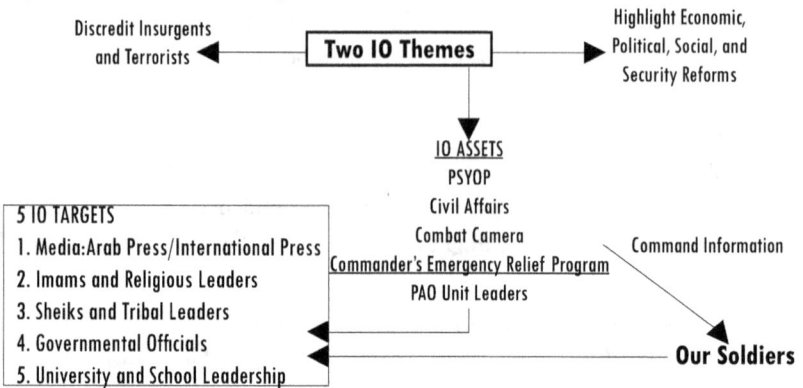

comfort zones, personal involvement is especially necessary to ensure that the commander's concept is executed according to plan.

After establishing an initial IO cell, we obviously needed to develop an IO concept of operation that would synchronize our collective efforts. The centerpiece of this concept was the decision to dedicate brigade IO efforts toward two major themes and five target audiences (figure 3). The two major themes were to convince the silent majority of Iraqis in our AO that the economic, political, and social reforms being implemented were in their personal and national interest to support, and to discredit insurgent and terrorist activities in order to deny them support by the silent majority.

Our overall target audience was clearly the silent majority. However, to reach them and to ensure that our messages and themes would resonate with them, we determined that we needed to use mainly Iraqi proxies to convey our messages. We therefore identified five groups of Iraqis that had significant influence among the population: local imams and priests, local and district council members, staff and faculty from the universities, Arab and international media, and local sheiks and tribal leaders. Armed with a conceptual framework for conducting information operations throughout the brigade, we then wrote and published an IO annex. This leads to my sixth IO observation: An IO campaign has a greater likelihood of success if messages are simple and few, and repeated often.

Repeating themes and messages

While developing my commander's guidance, I recalled that the average person has a hard time remembering even simple concepts if he is only exposed to the concept once. A person watching commercials on TV, for example, must watch the same commercial ten or twelve times before he retains the message and becomes inclined to buy the product. Keeping this in mind, we strove for sufficient repetition whenever we disseminated information. To influence the population, it was important to develop and repeat the messages that focused on our two themes, and to ensure that they were accurate and consistent.

Staying focused

Our ultimate IO objective was to convince the majority of the Iraqis in our area that they should tolerate our short-term occupation because we, working with them, could create conditions that would lead to a better life for them individually and collectively. As mentioned earlier, we developed two overarching themes that, if communicated often and convincingly to the Iraqis, would contribute to our goal. To support our first theme (convincing the Iraqis that it was in their personal and national interest to support reform initiatives), we defined success as progress being made economically, socially, politically, and in security. To support our second theme (discrediting the insurgents and terrorists), we took every opportunity to draw attention to the destructive, vicious disregard the enemy had for the Iraqi people and the adverse effects their actions were having on individual and national progress.

With much command emphasis, we developed metrics and the information requirements to support them. We then meticulously collected information from throughout the brigade area in support of the metrics, which we integrated into IO messages to bolster our two major themes. Using "economic reform," for example, we tracked the status of every brigade renovation and reconstruction project. These projects were effective in supporting our first theme because they directly resulted in quality-of-life improvements for the Iraqis. Better schools, cleaner drinking water, functional sewage disposal, more efficient distribution of electricity in our area, functioning health clinics and hospitals, and repair of university schools are some examples of the information we used to substantiate our claims.

We maintained a running total of the new projects we had started, how many were in various stages of completion, how many had been completed, and how much money the Iraqi transitional government, the U.S. Government, or the international community had contributed to each. We also collected detailed information about insurgent and terrorist activities in our area to support our second theme. We tracked the number of Iraqi citizens killed or injured because of insurgent activities each day, the type of property damage and associated dollar value of damage caused by the insurgents, and the adverse effect that insurgent attacks were having on the quality of life (hours of daily electricity diminished, fuel shortages, number of days lost on completing vital infrastructure projects, and so forth).

One of our early IO challenges was maintaining consistent, accurate, noncontradictory IO messages. To address the challenge, we codified in our IO annex the kind of information to be collected, along with the requirement to roll up such information and submit it to the brigade IO cell each week. The cell used this precise, accurate information to develop talking points for all brigade leaders, and the points were disseminated to subordinate commands in our weekly fragmentary order. As a result, when we spoke with the media, government officials, imams and priests, university staff and faculty, and tribal sheiks, we were all saying the same thing – one band, one sound – all the time, with talking points crafted to reinforce our two themes.

Making IO part of overall operations

Because battalion leaders were busy fighting a war and dealing with lots of other problems, it would have been easy for them to place less and less priority on the brigade IO plan until it was subsumed by some other priority. Therefore, I knew that if I did not emphasize IO, it would not become a cornerstone of our daily operations. I felt strongly enough about the need for a brigade-wide IO effort that I made it one of my top priorities, so that the battalion commanders would follow suit as well.

Almost all of our IO activities were codified in our IO annex, which we developed and issued as a fragmentary order. This detailed annex described our two major themes and five target audiences, and it directed subordinate commands to conduct meetings, either weekly or bi-weekly depending on the audience, with the leaders of our targeted audiences (figure 4). The annex also directed subordinate commands to collect the information needed to support our weekly talking points, provided specific guidance on how to work with the media, and stipulated many other tasks that were necessary to support the brigade IO concept. I did not leave the "who and how often" up to the battalion commanders. They could not say, "I know I'm supposed to meet with these imams this week, but I'm just too busy." The engagement was required.

To manage this process further, I required weekly reports. If a

Figure 4. 2BCT IO battle rhythm.

	Sunday	Monday	Tuesday	Wednesday	Thursday	Friday	Saturday
2BCT	O&I 0900–0930	O&I 0900–0930	O&I 0900–0930	CMD & STAFF 0900–1000	O&I 0900–0930	O&I 0900–0930	O&I 0900–0930
	TF Weekly IO Summary Due 1200	IOWG 1100–1200	Weekly FRAGO Brief 1000–1200	NGO-IGO Meeting 1300–1500	R&S Backbrief 1000–1030	TF CDR IO Backbrief* 1230–1400	BDE Tng Mtg 0800–1000
	Targeting Board 1300–1430	QOL/LRR 1100–1300	BI-WEEKLY PRESS CONF. 1400–1500	RELIGIOUS LDR/SHEIK MEETING** 1300–1430	LRR 1100–1300	IOWG w/CDR** 1430–1530	S2 Mtg 0800–1000
	Local Govt. 1600–1200			Media Contact Report Due 1600	Media Contact Report Due 1600	Media Contact Report Due 1600	CDR's Day 1100–1400
	Media Contact Report Due 1600	Media Contact Report Due 1600	Media Contact Report Due 1600	S5/CA 1900–2100	Eng Project Meeting 1900–2000	TGT Noms Due 1800	Media Contact Report Due 1600
							Local Govt. 1600–1900
1AD						G5/S5 Meeting 1000–1200	

■ STRIKER 6 attends black shaded box events   VER: 25 Jan (CMTC)

* TF CDR IO Backbrief – Last Friday of month   ** EVERY OTHER WEEK

IGO, Intergovernmental Organization; O&I, Operations and Intelligence; QOL, Quality of Life; LRR, Logistical Readiness Review

commander failed to conduct a mandatory target audience engagement, I demanded an immediate justification. I do not typically operate in such a directive mode, but I felt such an approach was necessary, at least initially, to ensure that our IO plan developed into something more than a good idea.

Not surprisingly, there were some growing pains, even gnashing of teeth. But once commanders saw and felt the positive effects we were having, they bought in and the program became a standard part of how we did business.

To institutionalize the IO process even further and to habituate battalion commanders to it, I required monthly backbriefs, not unlike quarterly training briefings but focused on IO activities. The commanders briefed from prepared slides in a standardized format. They addressed such topics as the frequency of engagements with targeted audiences in their areas, the number of Arab press engagements conducted, and a roll-up of directed information requirements collected that month in support of our major IO themes. They were also expected to brief what they had accomplished for the month, and what their plans were for the next month, specifically highlighting planned changes and adaptations.

This briefing technique improved my situational awareness of the brigade's IO and provided a forum where leaders could share ideas and best practices. For example, one of the commanders might brief a new way in which insurgents were attempting to discredit Coalition forces, then address what he was doing to counter it. Other commanders could anticipate similar attempts in their AOs and take proactive measures to deny insurgent success. When we executed more traditional operations, I gave the battalion leadership great latitude to plan and execute in their battlespace. For information operations, however, I felt I had to be directive to ensure compliance with the plan I envisioned.

Developing talking points

We developed two sets of talking points to support our themes. The first set came from input the battalions provided weekly. It addressed what the insurgents were doing that adversely affected the Iraqis, and detailed actions showing how Iraqi lives were getting better because of cooperative Coalition and Iraqi successes. This information was consolidated and vetted by the IO cell, then pushed back out to the battalions to provide consistent, accurate talking points and to preclude us from committing IO fratricide by contradicting ourselves.

The other set of talking points were templated standing sound bites for engagements of opportunity that might occur due to catastrophic events. We could not predict when, but we knew suicide bombings and other sensational insurgent attacks were going to occur, and we wanted officers who would be the first to arrive to have some handy formatted guidance with which to engage the media and local officials who were sure to show up. These standard talking points gave the first company commander or battalion commander on the scene sufficient material to talk to the media with confidence.

The talking points also helped commanders stay on theme and make the points that we wanted to make. While the talking points were general, they were still specific enough and timely enough to satisfy the press. The standard talking points also allowed us to shape the information environment somewhat by suggesting what the focus of an incident should be rather than leaving it up to the media to find an interpretation (which the insurgents were often clever at providing).

Along with the five target audiences that we engaged with our weekly talking points, we actually had a sixth audience: our own soldiers. As our own quality of life began to mature, our soldiers gained easy access to satellite TV. Typically, they would watch CNN, the BBC, FOX, or some other major international news media. It quickly became clear to us that if these organizations were the most influential sources of information soldiers were exposed to, they would receive unbalanced information from which to develop their opinions of the effect their efforts were having in this war.

I remembered talking about soldier morale with Major General Martin E. Dempsey, who said that a soldier's morale was a function of three things: believing in what he is doing, knowing when he is going home, and believing that he is winning. Watching the international news was not necessarily going to convince anyone that we were winning. Therefore, we decided to take the same information we were collecting to support our two IO themes and use it as command information for our soldiers, so they could better understand how we were measuring success and winning, and be able to appreciate the importance of their contributions.

Value of societal and cultural leaders

For communicating our message to the Iraqis, our challenge was twofold: we had to exhaust every means available to ensure the Iraqis heard our messages, and (frankly the greater challenge) we had to get them to believe our messages. We constantly strove to earn the trust and confidence of the Iraqis in our area by consistently being truthful with them and following through on our word. Many if not most of the Iraqis we were trying to influence with our IO themes did not have access to us, did not have an opportunity to change their opinions about our intentions, and tended not to believe anything a Westerner said to them. For our information to resonate with the population, we realized we had to reach the most trusted, most influential community members: the societal and cultural leaders.

We hoped to convince them to be our interlocutors with the silent majority. We identified the key leaders in our AO who wielded the greatest influence. These included clerics (Sunni and Shi'ite imams and Christian priests from Eastern Orthodox churches), sheiks and tribal leaders, staff and faculty at the universities (a group that has incredible influence over the young minds of college-age students), local government officials whom we were mentoring, and finally, select Arab media correspondents.

We began our leader engagement strategy by contacting members of local governments at neighborhood, district, and city council meetings. We sat side by side with elected local council leaders and helped them develop their democratic council systems. Eventually, we took a backseat and became

mere observers. My commanders and I used these occasions to cultivate relationships with the leaders and to deliver our talking points (never missing
an opportunity to communicate our two brigade themes). We typically met
weekly or bi-weekly with prominent religious leaders, tribal sheiks, and university staff and faculty to listen to concerns and advice and to communicate
the messages that supported our IO themes.

The meetings were excellent venues for our target audiences to express whatever views they were willing to share. Usually, we initiated a session
with them by asking "What are we doing that you think is going well in your
neighborhoods? What are we doing that is not going so well?" Not unexpectedly, 95 percent of their comments focused on what we were not doing so
well (from their point of view). But this dialogue, however negative the feedback might have been, gave them a forum to communicate to us the rumors
they had heard through the Iraqi grapevine. In turn, this gave us a platform
to counter rumors or accusations and, using the detailed information we had
collected, to invalidate untrue or unsubstantiated rumors or allegations. After
fostering relationships with the leaders from our target audiences over a period of time, we were able to refute anti-Coalition rumors and allegations with
some degree of success.

These venues also gave brigade leaders insights to follow up on any
allegations of unacceptable actions by any of our units or soldiers. In fact,
when any group raised a credible point that involved something I could affect,
I tried to act on it immediately. In our next meeting with the Iraqi leaders,
I would explain to them what I had discovered based on their allegations and
what I was doing about it. For example, a sheik alleged that we were intentionally insulting Arab men when we conducted raids. He specifically referred
to our technique of placing a sandbag over the head of a suspect once we apprehended him. I told him that doing so was a procedure we had been trained
to perform, probably to prevent prisoners from knowing where they were
being held captive. His response was that everybody already knew where we
took prisoners and that it was humiliating for an Iraqi man to be taken captive
in his house and have "that bag" put on his head, especially in front of his family.
The sheik's point was that by following our standard operating procedure to
secure prisoners, we were creating conditions that could potentially contribute to the insurgency.

Back at headquarters we talked this over. Why do we put bags on
their heads? Nobody had a good answer. What do we lose if we don't use the
bags? What do we gain if we don't? We decided to discontinue the practice.

Whether doing so had a measurable effect or not is unknown, but the change played well with the target audience because it was a clear example that we valued the people's opinions and would correct a problem if we knew about it. This simple act encouraged the people to share ideas with us on how we should operate and allowed them to say, "See, I have influence with the Americans." This was useful because it stimulated more extensive and better future dialog. Another benefit of these engagement sessions was an increase in our understanding of the culture. We had not undergone cultural training before deploying to Iraq, but we received a significant amount of it through on-the-job training during these sessions. In fact, many of the tactics, techniques, and procedures we adopted that allowed us to strike a balance between conducting operations and being culturally sensitive came from ideas presented to us during meetings with leaders of our key target audiences.

Embedded media

Everybody thinks embedded media is a great concept. I do. I had James Kitfield from the *National Journal* embedded in my unit for three months during my tour in Iraq. That is an embed – somebody who stays with the unit long enough to understand the context of what is going on around them and to develop an informed opinion before printing a story. Unfortunately, as Phase IV of the operation in Iraq began, the definition of what an embed was for some reason changed to mean hosting a reporter for three or four days or even just one day. That is risky business because a reporter cannot learn about or understand the context of the issues soldiers face and, consequently, has a greater propensity to misinterpret events and draw inaccurate conclusions. Realizing this, I made it a brigade policy that we would not allow reporters to live with us in the brigade unless they were going to come down for an extended period of time.

Reporters who wanted to visit us for a day or two were welcome, but they had to go home every night because I was not going to expose them to, or give them, the same kind of access a true embed received if they did not want to invest the time needed to develop a sophisticated understanding of the environment the soldiers faced, the decisions we were making, and the context in which we were fighting. Therefore, my seventh IO observation is that reporters must earn their access.

Unfortunately, it is also my experience that some reporters come with a predetermined agenda and only want to gather information to support some particular political or personal slant for a story they are already developing. However, I learned by experience who those reporters were and what to expect from them. No matter what we do, we are not going to change some reporters' or publications' mindsets. The best way to work around a biased and unprofessional journalist is by being more professional than they are and by developing a plan to deal with them.

Arab versus international media

Although the international press is an integral component of our IO effort, it was not our top media priority. While higher headquarters viewed U.S. and international media as their main media targets, our priority was more parochial: we regarded the Iraqi and Arab media as our main targets. As a result, most of the time I spent on the media was focused on the Arab press because it informed the population in my area. What most people were viewing on their new satellite TV dishes was Al Arabiya and Al Jazeera, not CNN, the BBC, NBC, or FOX. From my perspective, I was competing with the insurgents for the opinion of the silent majority, the wavering mass of Iraqi citizens who were undecided in whom they supported and who constituted the most important audience we needed to influence.

Weekly roundtables

The most effective technique we developed to engage the key members of the Arab press routinely was the bi-weekly, brigade-level news huddle. Since policy at that time did not permit us to conduct press conferences, we held small roundtables, something like the exclusive U.S. Department of Defense (DoD) press roundtables conducted in Washington, D.C. We allowed only the Arab press to come to these sessions; CNN, the BBC, and other international media were excluded. The Arab media was our target audience because it was our conduit of information back to the Arab community.

Every two weeks I invited Arab media representatives to my headquarters. In preparation, one of my PAOs drafted talking points and a script. I began each meeting with scripted comments emphasizing messages related to our two primary IO themes, then opened the floor to questions.

To focus our efforts and to determine from which venues the Iraqis received their news, we conducted surveys and ascertained which newspapers were read and which TV programs were watched in our battlespace. We then hired two Iraqis to be brigade press agents. Their main jobs were to facilitate attendance at our press roundtables and to promote the publication of our messages. They would go out, visit with various newspapers, and invite reporters to our press conferences. Typically, the press agents described how we conducted our press conference, provided reporters with the location and frequency of our meetings, and coordinated the reporters' clearance for entry into our forward operating base. Finally, the press agents would stress to the reporters that they were not only allowed but encouraged to ask anything they wanted.

It was not unusual to have anywhere from eight to ten newspaper reporters attend these meetings, among them representatives from Al Jazeera, Al Arabiya, and one of the Lebanese satellite TV stations. After the press huddle I usually did offline interviews with the Arab satellite stations.

Engaging Al Jazeera and Al Arabiya

Al Jazeera and Al Arabiya, for the most part, enjoy a justifiably bad reputation in the West because of their biased reporting style. But the fact is they report to the audience we need to influence, so why not develop a rapport with them so that maybe we can get some of our messages across to the Iraqi public?

When Al Jazeera reporters first came to one of our press huddles, they were distant. However, after three or four meetings they began warming up to us and later, they became just as friendly as any of the other reporters attending. We can, if we put enough effort into it, develop a good working relationship with almost any reporter as long as we are truthful and honest. They cannot help but respect us for that and, much of the time, respect is rewarded with fairer and more balanced news accounts because reporters know they can trust what we are saying. It is a mistake not to allow Al Jazeera and other Arab media access simply because we do not like much of what they report. We need to work with them specifically if we want more accuracy and balance. We cannot just censor them, deny them access, or fail to respect them because, ultimately, they talk to Arab peoples in their own language and are the most likely to be believed. Not to engage them or work with them is to miss tactical and strategic opportunities.

Handbills

Another important tool in our efforts to communicate IO themes to the Iraqi public was handbills. Generally, we Westerners dismiss handbills as a trivial medium because we associate them with pizza advertising, close-out sales, and other such activities. In Iraq, hand-distributed material in the form of flyers and leaflets is an effective way to distribute IO messages.

To take the initiative away from the insurgents, we developed two different types of handbills: one to address situations we faced routinely (figure 5), another for mission-specific operations or incidents (figure 6). Standard handbills spread news about such events as improvised explosive device (IED) incidents, house raids, and road closings (usually to clear an IED). Because we wanted to ensure that we had a way to take our IO message straight to the local population as soon as an opportunity presented itself, every mounted patrol carried standard flyers in their vehicles at all times. Thus, when soldiers encountered a situation, they could react quickly.

We also relied on handbills tailored to specific incidents that had occurred or operations we were conducting. For example, we might draft a handbill addressing an insurgent incident that had killed or injured Iraqis citizens in a local neighborhood. Being able to produce and disseminate rapidly a handbill that exposed the callous and indiscriminate nature of insurgent or terrorist activities while a local community was reeling from the attack was powerful and effective.

When developing handbills, we followed two important guidelines:

Figure 5. House raid handbill.

We apologize for this inconvenience. We have been forced to conduct these types of search operations because people in your community have been attacking Iraqi and Coalition security forces.

Thank you for your cooperation.

Iraqi and Coaltion security forces are conducting operations to defeat terrorists who use your community to plan and execute attacks against Iraqi citizens and Coalition forces.

ensure that messages were accurately translated, and ensure that the handbills were distributed in a timely manner. Much careful, deliberate thought went into the scripting of our messages. We made sure our best interpreters translated the material, and we vetted each translation through multiple interpreters to ensure accuracy.

It is an unfortunate characteristic of war that tragedy invites the greatest interest in political or social messages. As a result, the best time to distribute a leaflet, as exploitative as it seems, was after an IED or some other sensational insurgent attack had resulted in injury or death. A population grieving over lost family members was emotionally susceptible to messages vilifying and condemning the insurgents.

Consequently, we would move rapidly to an incident site and start distributing preprinted leaflets to discredit the insurgents for causing indiscriminant collateral damage. We also requested help in finding the perpetrators of the attack. Such leaflets brought home immediately the message that the insurgents and terrorists were responsible for these events and that the best way to get justice was to tell us or the Iraqi security forces who the insurgents were and where they could be found. This technique, which helped drive a wedge between the insurgents and the locals, often resulted in actionable intelligence. Quick distribution of leaflets helped influence our popula-

Figure 6. Handbill addressing specific incident.

Anti-Iraqi forces are operating in your neighborhood.

25 casualties on 18 January:

13 Iraqis killed—including an 8 year old boy!!

12 severely wounded—11 Iraqis and 1 Coalition Soldier

Only you can help stop this violence. Report all IEDs and suspicious activity to Iraqi or Coalition Security Forces.

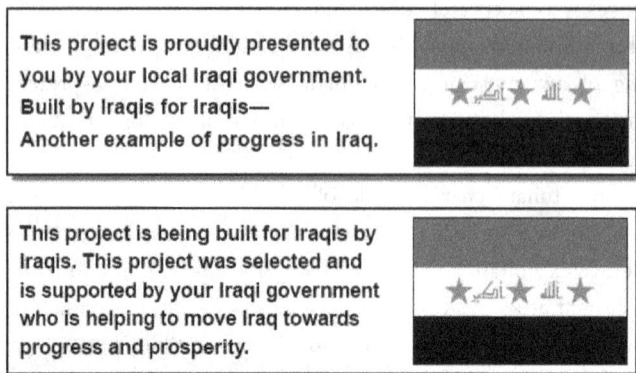

This project is proudly presented to you by your local Iraqi government. Built by Iraqis for Iraqis— Another example of progress in Iraq.

This project is being built for Iraqis by Iraqis. This project was selected and is supported by your Iraqi government who is helping to move Iraq towards progress and prosperity.

Figure 7. Iraqi success handbill.

tion before the insurgents could spin the incident against us.

We also drafted handbills that informed the Iraqis about local or national infrastructure progress (figure 7). We highlighted successes, such as the increased production of electricity in the country and improvements in the amount of oil produced and exported. We specifically designed these leaflets to convince the population that progress was occurring.

Measures of IO effectiveness

As with all operations, gauging IO effectiveness is important; however, the process of measuring IO success is not a precise science. That noted, we did discover certain simple techniques to identify indicators that we found useful for measuring effectiveness.

Iraqi PAOs

Iraqi PAOs were indispensable to our success with the Iraqi and Arab press. They were instrumental in soliciting Arab media correspondents to attend our bi-weekly brigade news huddles and in gauging what was being published or broadcast that directly affected our area of operations.

We hired two Iraqi interpreters and dedicated them to 24-hour monitoring of Arab satellite news. That's all they did: they watched satellite news television in our headquarters and noted every story that was aired about operations in Iraq. Through their efforts we were able to determine that our information operations were having the intended effect because of an increase in the number of accurate, positive stories published or aired in local papers and on satellite TV.

Updates and analysis from this monitoring process became a key part of the daily battle update brief. The PAO briefed us on newspaper articles or Arab TV stories related to our operations. For example, a story might have appeared on Al Jazeera about some particular issue or event in the brigade AO that might have been incorrectly reported. We would respond by developing an IO action to counter the story. This type of monitoring told us about the type of information being directed at the local population, which in turn allowed us to take action to counter or exploit the information.

Lack of adverse publicity

A similar key indicator that our IO efforts were succeeding was a lack of adverse publicity. While we were in Baghdad we raided eight mosques, but received no adverse publicity other than from a few disgruntled imams. To our knowledge, these raids were not reported by either the Arab or the international press. Nor did these raids prove to be problematic in feedback from the various target audiences we were trying to influence. We attributed this success to the meticulous IO planning we did for every sensitive site we raided. Ultimately, we developed a brigade SOP that detailed the IO activities we were required to do before, during, and after such raids.

Increase in intelligence tips

Another indicator of success was the increased number of intelligence tips we received. We determined that there was a correlation between the number of tips we received from unpaid walk-in informants and the local population's growing belief that they should distance themselves from the insurgents and align themselves with Coalition reform efforts. By comparing week after week how often local citizens approached our soldiers and told them where IEDs were implanted or where they were being made, we had a pretty good idea that our efforts to separate the insurgents from the population were working.

The wave factor and graffiti

An informal but important indicator was what we called the wave factor. If you drive through a neighborhood and everyone is waving, that is good news. If you drive through a neighborhood and only the children are waving, that is a good but not great indicator. If you drive through a neighborhood and no one is waving, then you have some serious image problems.

A similar informal indicator was the increase or decrease of anti-Coalition graffiti.

Monitoring mosque sermons

A more sophisticated indicator came from reports of what had been said at mosque sermons. Monitoring imam rhetoric proved to be an important technique because messages delivered during sermons indicated whether or not imams were toning down their anti-Coalition rhetoric. If they were, we could claim success for our program of religious leader engagements. Feedback on what was said inside the mosque steered us to those imams we specifically needed to engage. For example, I would be briefed that a certain imam was still advocating violence against Coalition forces or that he was simply communicating false information. We would then tailor our IO efforts to engage that particular imam or other local neighborhood leaders so that he might modify his behavior and rhetoric.

The way ahead

In Iraq's COIN environment, information operations are important tools for achieving success. I believe the program we developed, with its focus on engendering tolerance for our presence and willingness to cooperate (rather than winning hearts and minds), and its basis in consistent, reliable actions supported by targeted communications to specific audiences, paid dividends.

Repetition of message, accuracy of information, and speed of delivery were key to executing our plan. Ultimately, those of us tasked with counterinsurgency must always keep in mind that we are really competing with the insurgents for influence with the indigenous population. In Iraq, that means convincing the population that they should tolerate our short-term presence so that economic, political, social, and security reforms can take root and ultimately give them a better country and a better life. To achieve this goal, we must dominate the IO environment. To dominate the IO environment, we need to ensure that information operations receive the same level of emphasis and involvement that our commanders have traditionally allocated to conventional maneuver operations. Until our Army matures in its development of doctrine and approach to training for insurgencies, commanders at all levels will need to play a prominent role in developing, implementing, and directing IO within their areas of operation.

One of the many strengths our Army enjoys is that it is an adaptive, learning organization. Significant changes are already taking place as we begin to learn from the lessons of fighting an insurgency. Our Combat Training Centers are implementing changes to their training models to better integrate IO into rotation scenarios. Their challenge will be to give rotating forces an irregular warfare experience that acknowledges and rewards good IO planning and execution by our soldiers. The addition of IO, PA, and CA officers, PSYOP NCOs, and PAOs to maneuver brigades is encouraging, and the offering of COIN electives at the Command and General Staff College (CGSC) indicates real progress. However, there is still more to be done before our soldiers and our Army can comfortably employ IO as a key instrument for waging war against an irregular enemy. Some of the following suggestions are already being considered and will soon be implemented; others I hope will spark some debate as to their merits:

•   Do more than add a COIN elective to the CGSC curriculum. Immediately require COIN instruction at all levels in our institutional training base.

•   Integrate cultural awareness training as a standard component in our institutional training base curriculum.

•   Increase the quality and quantity of media training provided to Soldiers and leaders.

•   Consider compensating culture experts commensurate with their expertise.

•   Reassess policies and regulations that inhibit our tactical units' ability to compete in an IO environment. The global communications network facilitates the near-instantaneous transmission of information to local and international audiences, and it is inexpensive and easy to access. Our Soldiers must be permitted to beat the insurgents to the IO punch.

Essential IO observations

1.   To be effective, tailor themes and messages to specific audiences.

2.   You have no influence over the press if you do not talk to them.

3.   There is a direct correlation between your credibility and your ability to improve demonstrably the quality of life, physical security, and stability in a society.

4.   A major IO goal at tactical and operational levels is getting the citizens in your AO to have trust and confidence in you.

5.   The commander's vision and intent are essential, but when you direct

subordinates to perform outside their comfort zones, your personal involvement is especially important to ensure that your concept is executed according
to plan.

6.   An IO campaign has a greater likelihood of success if messages are
simple, few, and repeated often.

7.   Reporters must earn their access.

In closing, the model of information operations I have advocated here
is simply one way to conduct IO at brigade level and below.   This model is
not intended to be the only way.   The unique aspects of each operational
environment, our national goals in wartime, the culture of the indigenous
population, and many other factors will ultimately dictate each commander's
concept of information operations. The important thing is to develop a plan
and to execute it aggressively. Failing to do so will give the insurgent a perhaps
insurmountable advantage.

15

Humint-Centric Operations: Developing Actionable Intelligence in the Urban Counterinsurgency Environment

BG Ralph O. Baker, USA

This article was solicited from the author by Military Review *as a companion piece to his article, "The Decisive Weapon: A Brigade Combat Team Commander's Perspective on Information Operations," published in May-June 2006 (see chapter 13). It is based on an unclassified briefing BG Baker presents regularly to leaders preparing to deploy to Iraq and Afghanistan.*

A few weeks after assuming command of the 2d Brigade Combat Team (2BCT), 1st Armored Division, I found myself sitting in a tactical command center in downtown Baghdad conducting a brigade cordon-and-search. The reports flooding in from my battalion commanders were virtually all the same: "STRIKER6, this is REGULAR6. Objectives 27, 28, 29 secure and cleared. Nothing significant to report. Over." We spent nearly ten hours searching for insurgents and weapons in hundreds of dwellings throughout our objective area, a bad neighborhood off Haifa Street that was a hub of insurgent activity – and for what? Ultimately, we captured a dozen weapons and a handful of suspects.

Much more worrisome to me than the meager results of our operation was the ill will and anger we had created among the Iraqi citizens who were the unwelcome recipients of our dead-of-night operations. I had been on enough such sweeps already to picture the scene clearly: mothers crying, children screaming, husbands humiliated. No matter how professionally you executed such searches, the net result was inevitably ugly.

That profoundly disappointing experience led me to a blunt realization: our dependency on conventional intelligence collection methods and our failure to understand the negative perceptions our actions were generating among Iraqi citizens threatened to doom our mission. If we did not change our methods, and change them quickly, we were not going to be successful in the urban counterinsurgency (COIN) environment in which we found ourselves. As a result of that realization, I made two decisions in the ensuing days that affected the way our combat team would operate for the remainder of our deployment. First, we would reform the way we conducted intelligence operations, and second, we would make information operations (IO) a pillar of our daily operational framework.

My purpose in writing this article is to share with the reader insights and lessons learned from the reform of our intelligence operations; specifically, what we learned by conducting human intelligence (HUMINT)-centric operations in a heavy BCT in Iraq. To that end, I want to describe briefly the initial state of my BCT and our area of operations (AO), identify the major intelligence challenges that we faced, and offer solutions and techniques we adapted or developed in order to overcome our challenges.

Background

Second BCT deployed to Iraq in May 2003. We were a conventional heavy BCT, task-organized with two mechanized infantry battalions, a cavalry squadron, an armor battalion, a field artillery battalion, an engineer battalion, a support battalion, and a military police battalion. The BCT's train-up prior to deployment had focused on conventional, mid- to high-intensity combat, and our battalion and brigade headquarters and staff processes were still optimized to fight a conventional threat.

Our AO included two districts in Baghdad – Karkh and Karada. Within these two districts lived somewhere between 700,000 and a million citizens, among them Sunnis, Shi'ites, and the city's largest population of Christians. Our AO also included the heavily fortified Green Zone and several neighborhoods with large populations of retired Iraqi generals, plus numerous ethnic, sectarian and political entities (either preexisting or emerging, such as the Supreme Council for the Islamic Revolution, the Islamic Dawa Party, and the Patriotic Union of Kurdistan).

With the exception of our counterintelligence warrant officer and a few other officers who had some previous exposure to HUMINT operations, we neither understood nor anticipated the inadequacy of our conventionally designed intelligence collection and analysis system. More importantly, almost no one understood the dominant role that HUMINT operations would play in developing actionable intelligence on a burgeoning insurgency.

The intelligence system we brought to Iraq was designed to identify conventional enemy formations, and our intelligence personnel were trained to conduct predictive analysis about an enemy based upon our knowledge of his equipment and doctrine. Exactly none of these conditions existed after Saddam's army was defeated.

Instead, we found ourselves in the midst of an insurgency, confronted by an elusive enemy force that wore no uniform and blended seamlessly into the local population. Conventional intelligence collection systems just don't work in this type of environment; our imagery operations, electronic reconnaissance, and standard combat patrols and surveillance operations were simply ineffective. After faithfully applying these conventional ISR (intelligence, surveillance, and reconnaissance) methods and assets to our combat operations, we netted almost no actionable intelligence.

Challenges

Realizing that we were fighting a growing insurgency and that the current conventional organization and training of our battalion and brigade intelligence sections were inadequate to address our needs, I decided to transition our conventional BCT intelligence system into a HUMINT-centric system. Not unexpectedly, a change of this magnitude for a unit engaged in combat against a growing insurgency presented many challenges. After considering the circumstances we faced in our AO and our leadership's lack of experience and familiarity with COIN operations, I found that our challenges could be grouped into three general categories: leadership, organization, and training.

Leadership

When people are confronted with substantive change that runs counter to their doctrine and training, it's natural for them to be uncomfortable and therefore hesitant to embrace that change. I assumed this would be the case from the beginning; thus, I set about implementing mechanisms to ensure that compliance with our intelligence changes was rapid and "as directed." From the beginning, I felt it was necessary to convince my commanders and staffs that transitioning to a HUMINT-based approach to intelligence was my absolute highest priority.

As a commander, you must set the conditions to ensure that your subordinates make HUMINT operations a priority and that they synchronize such operations with your headquarters. You must start out by providing a sound concept your subordinates can understand and follow: visualize the plan, describe it to your people, and then direct them in execution. After close consultation with my staff and other individuals with COIN experience, I presented a vision and draft organization for how I wanted units in the BCT to conduct intelligence operations. Central to our new intelligence system was the development of an extensive network of Iraqi informants. I felt it was absolutely key to identify and develop indigenous sources who had the ability to infiltrate Iraqi society and blend in. Such human sources of intelligence represent a critical capability that no ISR technology, no matter how sophisticated or advanced, can match.

Once we had decided to rely primarily upon informants for our intelligence collection, we modified our analysis process to bring it more in line with police procedures. This meant a heavy reliance on evidentiary-based link diagrams to associate individuals with enemy cells and networks, and some conventional pattern analysis when appropriate. Units were also directed to modify the organizational structures of their intelligence sections to accommodate new functional requirements such as intelligence exploitation cells, more robust current operations and plans cells, and additional subject matter experts who could support analysis and exploitation activities.

After we developed a concept and described it to the BCT's leaders, the final (and most leader-intensive) part of our transition was getting those leaders to buy in. I fully expected that many of my subordinate commanders would be very uncomfortable changing their intelligence organizations, collection assets, and analysis processes, particularly in the middle of a war. Throughout their careers, they and their Soldiers had experienced only conventional military intelligence operations. Forcing them to abandon a system they were comfortable with and that they thought adequate required commanders at all levels, starting at brigade, to stay personally involved in all aspects of the transformation.

HUMINT battle rhythm

Anticipating that I would likely face some resistance from within my organization, I implemented mechanisms that would allow me to promote compliance, conformity, understanding, and confidence in our new approach to intelligence collection and analysis. Two particularly useful venues that allowed me to stay personally involved in intelligence operations with my subordinate leaders were weekly reconnaissance and surveillance (R&S) back-briefs and BCT after-action reviews (AARs).

My weekly intelligence battle rhythm consisted of a brigade intelligence targeting meeting on Sunday, followed by a BCT fragmentary order on Tuesday, and then the R&S meeting on Thursday. I personally chaired the latter, with my intelligence officer (S2) and all the BCT's battalion operations officers (S3s) in attendance.

R&S meeting

The R&S meeting was particularly useful for several reasons. First, it allowed me to confirm that the decisions, priorities, and guidance I had provided during my weekly targeting board had been accurately disseminated and interpreted by my subordinate commands. Second, it allowed me to monitor our weekly recruitment and development of informants, who were absolutely central to our HUMINT-based intelligence program. Third, it gave me the opportunity to provide or clarify guidance directly from the weekly brigade intelligence FRAGO to all of the BCT S3s. Fourth, it improved my situational awareness of each of my battalion AOs. Finally, taking the time to chair this meeting personally demonstrated my commitment to making HUMINT-centric operations a top priority in the BCT.

During these meetings, the battalion S3s were required to brief me on a number of mandated topics: the priority of their collection actions, the status of informant recruitment and training, the allocation of intelligence collection assets, and any additional R&S support they required from brigade level or higher. Each battalion used a brigade-standardized matrix to crosswalk their priority intelligence requirements (PIR) with the asset or assets they planned to dedicate against their PIR. Any informant a battalion was using was listed on this matrix along with our organic collection assets.

The gathering of battalion S3s was one of our most important and productive intelligence meetings. It allowed me to assess the development and use of HUMINT assets, to ensure that the battalions' intelligence and collection requirements were nested with the brigade's, and to see how the battalions were progressing in the development and use of informants. It also provided a venue for the battalions to share lessons learned about intelligence targeting and collection.

Weekly BCT AAR

Another meeting that facilitated professional and informative dialog and gave me an opportunity to provide guidance to my commanders on intelligence issues was our weekly BCTAAR. It was held on Saturday, with every battalion commander and S2 attending. Each AAR began with the brigade S2 providing a detailed intelligence update of the entire BCTAO, followed by a discussion to ensure that we all shared a common enemy picture. This forum also allowed for the dissemination of intelligence lessons learned and best practices, and it gave me an opportunity to identify challenges and seek solutions from fellow commanders. Once our intelligence portion of the AAR was complete, the battalion S2s departed with the BCTS2 to synchronize BCT intelligence issues. Commanders stayed, and we continued our AAR of information and maneuver operations.

Net gain

These two weekly venues, the R&S meeting and the AAR, were essential to reforming our intelligence system and improving our individual and unit performance. They:

- Allowed me and the BCT S2 to emphasize or reinforce key components of our intelligence system routinely.
- Promoted a learning environment within a chaotic and fast-paced operational environment.
- Allowed the immediate sharing of lessons (good and bad) among key battalion leaders.
- Provided me with immediate feedback on how well we were adapting to our new system.
- Fostered a better understanding of, and leader buy-in to, our new method of intelligence operations.

Eventually, once leaders at all levels understood the new system of intelligence collection and analysis better, had gained experience with it, and had bought into it, I was able to back off and be less directive. My subordinate leaders were then free to adapt and modify their intelligence operations to fit best the needs of their AOs.

Organization and team building

It was relatively easy to visualize, describe, and modify the organizational structure and the processes that we adopted to transform our intelligence operations. The greater challenge was manning our new model and training our soldiers and leaders to conduct HUMINT operations.

As you would expect of a learning institution, our Army is changing its organizational structures and doctrine to address many of the shortcomings that units experienced early on in Iraq. In fact, the intelligence section of today's BCT now includes an exploitation cell – a capability (and personnel) we didn't have just two years ago. In addition to these organizational and doctrinal improvements, BCTs now have more experienced leaders who understand the need to collect HUMINT in the current operating environment.

That said, manning is one of the challenges units encounter when they try to adapt their intelligence sections to HUMINT operations. HUMINT-centric operations are very manpower intensive—the amount of information that must be collected, analyzed, and synthesized to produce actionable intelligence can be overwhelming. Personnel needed for activities such as document and technical exploitation, interrogations, informant meetings, and plans and current operations present additional manpower challenges. As a result, commanders will find themselves undermanned when they have to staff their transformed intelligence activities according to the typical authorization for a conventional intelligence section. The number of authorized billets and Military Occupational Specialties (MOSs) is simply inadequate to conduct and sustain HUMINT-centric operations. To develop an effective brigade intelligence team, you will have to find additional personnel to man it.

One way to address this shortcoming is to screen and select non-intelligence-MOS Soldiers from your BCT who have the required skills: intellectual capacity, technical expertise, and a natural proclivity to contribute to your intelligence effort. We never hesitated to take soldiers out of other sections or units to resource our intelligence sections. We had more than enough combat power in our organizations to overmatch the enemy in Iraq; what we didn't have was the depth and knowledge in our intelligence sections to find the enemy in the first place. To fix that, we integrated infantry and armor soldiers, cooks, communications specialists, and mechanics into our brigade and battalion intelligence sections. Commanders might also look closely at any National Guard and Reserve units attached to them during deployment. Many of the soldiers in these units already have unique skill sets (e.g., law enforcement, finance, computers and telecommunications) that make them excellent choices to serve as intelligence augmentees.

Having to build and train our intelligence team during combat was hardly ideal. Fortunately, units today have the opportunity to reorganize and train their intelligence sections and systems at home station prior to deployment. When we redeployed to our home station, we endured the typical personnel chaos (soldiers changing station and leaving the service) that occurs in the wake of a long deployment. After the majority of our personnel turnover was over, we immediately set about building and training our intelligence sections in anticipation of the brigade's next deployment.

Working closely with the Combat Maneuver Training Center (CMTC) and 1st Armored Division Headquarters, we developed a HUMINT-centric pre-rotational training program to facilitate the early and progressive training of our new intelligence teams. The chief of the division's All-source Collection Element (ACE) and CMTC's scenario writers and leaders developed a detailed enemy situation and database that replicated an insurgent-terrorist activity, one that could fully exercise the BCT's intelligence units. The intelligence flow began six months prior to commencement of our maneuver training exercise, as our intelligence sections at home received a steady stream of notional intelligence reports, interrogation debriefings, and programmed meetings with HUMINT sources. Using the torrent of information generated by the division ACE and CMTC, our intelligence sections were able to sustain the intelligence processes and techniques that we had developed while previously deployed to Iraq.

With that pre-rotational data and information provided in advance, our intelligence teams were required to conduct analysis, build link diagrams and target folders, and produce other intelligence products that passed along the hard lessons learned during our first deployment. We also continued to run our weekly intelligence battle rhythm just like we had in Iraq. My staff would provide me with current intelligence updates, recommend changes or additions to our PIR, conduct current analysis of insurgent organizations in our AO, and suggest intelligence targeting priorities.

These pre-rotational intelligence activities supported three important goals: first, they allowed us to train our newly staffed intelligence teams throughout the BCT based upon lessons we had learned and processes we had developed in Iraq. Second, they enabled us to maximize our training experience when we finally deployed for our rotation – instead of spending valuable time learning undergraduate lessons at an expensive postgraduate training event, we were able to hit the ground running based upon actionable intelligence our sections had developed over the previous six months. Finally, and most importantly, they developed the confidence of the new soldiers and leaders in our intelligence sections.

Informants

As I stated earlier, leveraging informants as our principal intelligence-collection asset constituted a significant shift from the way most of us had ever operated. The theory and logic behind using local sources to obtain information and intelligence is easy to grasp; however, the practical aspects of developing these nonstandard collection assets are less obvious.

In general, we had two challenges with informants: finding them and training them. Initially we relied upon informants who routinely provided unsolicited information to our units. We would track the accuracy and consistency of the information they gave us and, after they established a credible and reliable track record, we would begin to reward them for useful information. Later on, as our knowledge of our AO improved and, more importantly, our understanding of the culture and the nuances of local demographics increased, we became more savvy and cultivated informants from different ethnic, sectarian, political, tribal, and other groups within our AO. Eventually, the brigade's intelligence sections developed a rapport with three to five informants who consistently provided reliable information we could develop into actionable intelligence.

Among our informants were members of political parties, local government officials, prostitutes, police officers, retired Iraqi generals, prominent businessmen, and expatriates. Of course we recognized that there was risk associated with using informants. For example, we were concerned that they might be collecting on us, or that the information they provided might have been designed to settle personal vendettas. Consequently, our BCTS2 and counterintelligence warrant officer developed a vetting program to minimize such risks. All of our informants were screened to validate the quality of their information and to check their motivations for providing it. We also implemented careful measures to ensure that informants were not collecting on U.S. forces or providing information that would put our soldiers at risk.

Once we determined that a potential informant was reliable and useful, it became necessary to train and equip him so that he could provide more accurate and timely information. We typically provided our informants with Global Positioning System (GPS) devices, digital cameras, and cell phones. The phones not only improved the timeliness of information, but also allowed informants to keep their distance from us, thus minimizing the chance they would be personally compromised. Later on, as Internet cafes began to flourish in the Iraqi economy, we helped our informants establish e-mail accounts and used that medium as another way to communicate with them.

GPS devices were also important, because most informants could not accurately determine or communicate address information that was sufficient to pinpoint target locations. With some basic training, our informants could use their GPSs to identify key locations using the military grid reference system. This increased the accuracy of location marking and measurably enhanced our ability to develop precise, actionable intelligence. Occasionally it was useful to give informants automobiles, too, to facilitate their movement and collection activities inside and outside our AO.

We discovered that identifying and training an informant was a complex and time-consuming process. Finding the right type of individual willing to work with you is both an art and a science. Our counterintelligence-trained Soldiers were instrumental in ensuring that we worked with the most reliable, most consistently accurate informants. Training and equipping our informants were key to their effectiveness and paid great dividends in terms of the volume and accuracy of their information. Because informants were the foundation of our HUMINT system in the brigade, we resourced them accordingly.

Collecting and exploiting evidence

Although developing indigenous sources of intelligence was central to the way we operated, we quickly discovered that there was another key component to our HUMINT-driven system: the collection and exploitation of evidence. It is not only frustrating, but also detrimental to your mission success to culminate an operation with the capture of insurgents or terrorists only to be directed to release them because your justification for detaining them can't endure the scrutiny of a military or civilian legal review. We quickly learned after a couple of very avoidable incidents that our ability to prosecute intelligence operations successfully was directly linked to the ability of our Soldiers to collect, preserve, and exploit evidence related to our captured suspects. To remedy that, we initiated a training program to give our soldiers and leaders the skills they needed to manage evidence.

Leveraging the experience and training of our military police, National Guardsmen with law enforcement skills, and FBI agents in country, we were able to rapidly train our soldiers on the essential requirements for capturing, securing, associating, safeguarding, and exploiting evidence. Once they were armed with this training and an effective HUMINT-based intelligence process, our seizure and detention rate for insurgents, terrorists, and other miscreants soared.

Closely linked to the collection and association of evidence to suspects was the exploitation of that evidence. Early in our deployment we were frustrated by the inability of organizations above brigade level to exploit evidence in a timely manner and then provide feedback that we could use.

This was particularly true when it came to captured computer hard drives and cell phones. The standard policy was that these items had to be expedited to division headquarters within 24 hours of capture. This made sense because division was the first echelon above brigade that had the knowledge and expertise to exploit these devices. Unfortunately, for many reasons the turnaround time to receive intelligence from echelons above brigade was typically too slow, or the resultant product too incomplete, to help us.

What we needed was the ability to exploit these items at the BCT level for tactical information, in parallel with the division and corps intelligence shops, which were focused on other priorities. Based upon our previous working relationship with the FBI team in country, we managed to get a copy of a software program the agency was using to exploit hard drives. My BCT-communications platoon loaded the software on their computers, received some basic training, and instantly we had the ability to exploit hard drives. We dedicated a couple of linguists to our communications platoon section, integrated this element into our S2X cell, and from then on conducted our own tactical-level technical exploitation of computers. We still had to forward hard drives and cell phones to division within 24 hours of capture, but now we just copied the hard drive, forwarded the complete captured system to division, and exploited the information simultaneously with the division.

This easy technical remedy to our hard-drive exploitation problem consistently provided big payoffs for us. The new capability was useful for documenting evidence to support the detention of an insurgent and for developing follow-up targets. We had the same challenge with cell phones. Unfortunately, we couldn't acquire the technical capability we needed to exploit them as we had with the hard drives. I believe that phone exploitation is yet another trainable skill and capability that we should give our BCT communications platoons.

As with cell phones and hard drives, we were challenged to exploit our detainees fully. Specifically, we had to get them to provide information, and then we had to exploit that information to incarcerate them or to assist us

in developing further intelligence to support future counterinsurgency operations. To address this challenge, we developed and adapted two useful tools as
we gained experience at tactical-level interrogations. One was a detailed line
of questioning that our HUMINT Collection Teams (HCTs) could use when
questioning detainees; the other was the "cage infiltrator" – an Iraqi informant
who would pose as a detainee in our holding facility to gather valuable intelligence from actual detainees.

Developed by the HCT team leader and the S2, a detailed line of
questioning is extremely important for prioritizing the avenues of questioning
that your trained and authorized interrogators pursue. It is an especially important tool given the latter's extraordinary workload and the limited amount
of time they can dedicate to initial and follow-up interrogation sessions.

As a commander, I found that it was imperative to take a personal
interest in the line of questioning our HCTs pursued. For example, it was important to ensure that their line of questioning meshed exactly with the BCT's
PIRs and intelligence targeting priorities. I spent a lot of time with my S2 and
battalion commanders refining our PIR and specific intelligence requirements
(SIR), reviewing and establishing collection priorities, and synchronizing our
collection efforts. This entire effort can be derailed if the line of questioning
your interrogators pursue isn't nested with your unit's priorities.

To ensure development of the most effective interrogation line of
questioning, my S2 required our HCTs to participate in the following five-step
process (weekly or mission-specific):

•   HCTs receive updated PIR and associated SIR from the unit S2.
•   HCTs receive a current intelligence briefing from the NCO in charge
of the unit S2X cell.
•   Senior HUMINT warrant officer attends the BCT commander's daily
intelligence briefings to facilitate his understanding of the latest changes in
intelligence priorities.
•   HCTs develop lines of questioning and back-brief the unit S2 and
senior HUMINT warrant officer.
•   HCTs conduct interrogations.

We found that it was easy for our HCTs to determine the right questions to ask as long as they thoroughly understood our current PIR and SIR
(which we continuously updated and refined).

Because detainees figured out very quickly that we treat prisoners
humanely, it was not long before many of them refused to provide useful information. During interrogations we would typically hear things like "I'm

innocent, I was just sleeping at my cousin's house when you arrested me," or "Saddam bad, Bush good, thank Allah for the USA." If we didn't have substantive evidence to link these detainees to a crime or insurgent activity, their strategy of denial, obsequious behavior, or happenstance alibi was difficult to dispute. One day, my S2 came to me with an idea. At his suggestion, we planted an informant in our holding facility with instructions to listen to the detainees' conversations and then report to us what they discussed. This technique, which we dubbed "cage infiltration," resulted in immediate intelligence.

Subsequently, we redesigned the individual spaces in our holding facility so that we could place our infiltrators in individual detention spaces, between suspected insurgent leaders and their possible followers. The only way these detainees could communicate among themselves was to talk past our infiltrator to their accomplice or cell member. Our interrogation teams would then remove our infiltrator under the guise of a routine interrogation, debrief him, and then return him to the holding area. Armed with the new information, our interrogators could often modify their line of questioning for more effective and productive follow-up interviews.

In a very short time, this technique became our single most effective method for gaining information and intelligence from our detainee population. An additional benefit to using cage infiltrators was that they were interactive. Over time, as they became more experienced and adept at what they were doing, they became quite clever at developing a dialog with their fellow detainees that would draw out additional information useful in incriminating the suspect or in developing future targetable information.

Another twist to this technique was the use of a taxi-driver informant. Despite our best efforts, there were times when we couldn't build a case strong enough to support the long-term detention of a suspect. When that happened, we would make our apologies for the inconvenience the suspect had endured and offer him a taxi ride back to his residence. It was not unusual for these suspects to brag to the driver or among themselves on their way home how they had deceived the "stupid" Americans. They would incriminate themselves in the process or reveal details that we could use to conduct follow-up COIN operations. Upon returning to our headquarters, the taxi driver was debriefed on the suspect's conversation. Based upon the nature of any new information the informant presented, we decided either to recapture the suspect or to cease pursuing him.

Ensuring that the line of questioning our HCTs pursued was nested with the BCT's intelligence priorities, coupled with some simple deception techniques such as using cage infiltrators in our holding facility, considerably improved the quantity and quality of intelligence that we obtained from our detainees.

Conclusion

Throughout the course of this article I have attempted to identify some of the major intelligence challenges my BCT faced during our first tour in Iraq. I have provided examples of how we met these challenges and adapted to best meet our needs at the time. I've also shared some of our more useful and effective practices in the hope that others may use or modify them to support their needs. I don't pretend that the examples and practices I've offered represent definitive solutions to the countless intelligence challenges units face in Iraq. My intent, rather, was to demonstrate that by direct and constant leadership involvement at all levels, conventional units can effectively organize, train for, and execute HUMINT-centric operations in a COIN environment with great success.

One final thought

This article is designed to complement a previous piece I wrote for *Military Review* ("The Decisive Weapon: A Brigade Combat Team Commander's Perspective on Information Operations") in which I described the contribution that IO made to our COIN efforts in Baghdad.[1]

Although HUMINT-centric operations and IO may appear distinctly different in terms of their aims, they are closely linked; in fact, they are mutually supportive. HUMINT-centric operations target the insurgent and the terrorist, but in doing so they produce precise and timely information that allows our Soldiers to locate and attack insurgent forces with surgical precision, minimum violence, and minor collateral damage. A corollary benefit is that our actions result in minimal harm and inconvenience to the local population, helping us to convince them that we have the intent and capacity to improve their security and daily lives by eliminating the insurgent threat.

Likewise, IO synergistically supports our intelligence efforts by convincing the local population that it is in their best interest, personally and nationally, to tolerate and even support our efforts to improve their lives.

---

1   Ralph O. Baker, "The Decisive Weapon: A Brigade Combat Team Commander's Perspective on Information Operations," *Military Review* 86 (May-June 2006): 33-51. Reproduced as chapter 13 in this anthology.

Through IO, we share with the population the progress that is being achieved politically, economically, and socially, and we ensure that they know about the violence and harm the insurgents are wreaking upon their fellow citizens and their nation.

Similarly, through IO we are able to let the population know that we can separate and protect them from insurgent-terrorist threats when they have the confidence to share targetable information with us. The more adept we become at conducting IO and influencing the population, the more information the population will provide to enable us to target the insurgents and terrorists. It's a win-win dynamic.

Given the environment our forces are operating in today and will continue to confront in the future, HUMINT-centric operations and IO are no longer merely "enablers" or supporting efforts. Quite simply, they are the decisive components of our strategy. Both of these critical operations must be embraced; they must become the twin pillars of the framework from which we operate. No longer can we allow our greater comfort with conventional combat operations to minimize these decisive components of a winning COIN strategy.

IV
Culture and Strategy

16

Transforming the Department of Defense Strategic Communication Strategy

COL Gregory Julian, USA

In the 2002 National Security Strategy (NSS) the Bush Administration identifies the need for a "different and more comprehensive approach to public information efforts that can help people around the world learn about and understand America." It also advocates "using effective public diplomacy to promote the free flow of information and ideas to kindle the hopes and aspirations of freedom of those in societies ruled by the sponsors of global terrorism."[1] This infers the need for a National Communication Strategy emanating from the President's administration, synchronized with communication strategies of all subordinate governmental organizations. The essence of strategic communication is to synchronize and coordinate public affairs, statesmanship, public diplomacy, and military information operations in concert with the actions of employing the national elements of power to achieve national objectives.

However, these ideas are not reflected in the subordinate strategies pertaining to the military. Within the Department of Defense (DoD); the 2005 National Defense Strategy (NDS) and the 2005 National Military Strategy (NMS) are devoid of any mention of communication strategy or military support to public diplomacy.[2] As a result, the DoD has not achieved its full potential in supporting Operations Enduring Freedom (OEF), Iraqi Freedom (OIF), and the overall Global War on Terrorism (GWOT) through strategic communication because of the lack of recognition for this essential capability. Even though this critical requirement was identified in the 2002 NSS,

---

1 George W. Bush, The National Security Strategy of the United States of America (Washington, D.C.: The White House, September 2002) p. 6, 31.

2 Donald H. Rumsfeld, The National Defense Strategy of the United States of America (Washington, D.C.: The Pentagon, March 2005).

and is widely recognized in subsequent government studies and reports as a significant shortcoming in the department's ability to achieve communication objectives – it remains an unfulfilled task.[3] The 2001 Defense Science Board's (DSB) report on Managed Information Dissemination and the 2004 Defense Science Board report on Strategic Communication both identify needed improvements in national and military communication efforts. Not until recently have specifications been spelled out for national and military strategic communication objectives for ongoing operations as they are now in the National Strategy for the Global War on Terror. The war on terror and supporting military operations have suffered from the absence of strategic communication strategy and creative approaches to competing in the global information environment. This deficiency is noted by Secretary Rumsfeld in his recent comments to the Council on Foreign Relations, "Our enemies have skillfully adapted to fighting wars in today's media age, but for the most part we, our country, our government has not."[4]

New and emerging technologies for delivering information to various audiences are not fully embraced by the military public affairs community. Some of the newer technologies are being explored and exploited not only by commercial news organizations but by adversaries in the GWOT while military public affairs lag behind. Various creative and successful initiatives are being explored by small segments of military public affairs, but the community as a whole is slow to adopt and expand the use of these new technologies.

Understanding various audiences and how they perceive our messages is essential to succeeding in supporting military operations through communication efforts. The diversity and fragmentation of the audiences, as well as the media, make for a challenging environment in which to compete. Some audiences may even shift sides depending on how a particular campaign affects their concerns. The world's media seem to be fixated on America's faults, and this makes the adversary's job easier by diverting attention from negative coverage of their actions to any error of the U.S. or the coalition.

The Defense Science Board Chairman wrote, "Effective strategic communication can prevent a crisis from developing and help diffuse a crisis after it has developed. To win in a global battle of ideas, a global strategy for

---

3 Defense Science Board, *Report of the Defense Science Board Task Force on Strategic Communication* (Washington, D.C. : Defense Science Board, 23 September 2004).

4 Ann Scott Tyson, "Rumsfeld Urges Using Media to Fight Terror," *Washington Post*, 18 February 2006, p. A07.

communicating those ideas is essential."[5] Many of the communication challenges we are experiencing during stability operations in Southwest Asia could
be mitigated by better synchronizing the DoD communications strategy verti-
cally and horizontally, and exploring new technology for reaching emerging
audiences.

This paper will examine DoD's Strategic Communication strategy
to support the Global War on Terror objectives. It will review the ways and
means that DoD employs to achieve its communication objectives, and evalu-
ate emerging information technologies to enhance future strategic communication requirements. It will also make recommendations to transform current
practices that will require reprioritizing resources in order to gain an advantage in the contemporary global information environment.

In the summer of 2004, the Deputy Secretary of Defense tasked the
Defense Science Board (DSB) to study strategic communications as part of a
study on the Transition to and from Hostilities. The DSB task force on strategic communication developed seven recommendations to improve the U.S.
ability to communicate with and influence worldwide audiences. The sixth
and seventh recommendations apply to the military and address the following:

Appoint the Under Secretary of Defense for Policy as the DoD focal
point for strategic communication and serve as the Department's principal on
the NSC's Strategic Communication Coordinating Committee.

The Task Force recommends that the Under Secretary of Defense for
Policy and the Joint Chiefs of Staff ensure that all military plans and operations
have appropriate strategic communication components, ensure collaboration
with the Department of State's diplomatic missions and with theater security cooperation plans; and extend U.S. STRATCOM's and U.S. SOCOM's
Information Operations responsibilities to include DoD support for public
diplomacy.

The Department should triple current resources (personnel and
funding) available to combatant commanders for DoD support to public diplomacy and reallocate Information Operations funding within U.S. STRAT-
COM for expanded support for strategic communication programs.

The National Military Strategic Plan for the War on Terrorism is the
first DoD plan since the release of the DSB report to address the requirement

---

5 Defense Science Board Chairman William Schneider, Jr., "Final Report of the Defense Science Board (DSB) Task Force on Strategic Communication," memorandum for Acting Under
Secretary of Defense, Washington, D.C., 23 September 2004.

for strategic communication components.[6] The DoD's "strategic communication objectives in the GWOT are to align Coalition and partner nations against violent extremism, provide support for moderate voices, dissuade enablers and supporters of extremists, deter and disrupt terrorist acts, and counter ideological support for terrorism."[7] A sustained, proactive communication effort will be required to meet these objectives. The plan calls for combatant commanders to tie in all strategic communications capabilities and synchronize them with interagency efforts to achieve synergistic communications effects. "An important change the government needs to make is to incorporate communications planning into every aspect of the war on terror," Rumsfeld said.[8]

In recent months the Office of the Assistant Secretary of Defense (Public Affairs) (ASD (PA)) established a position for a Deputy Assistant Secretary of Defense (Joint Communications) (DASD (JC)) to shape DoD-wide communication processes and manage the public affairs transformation program.[9] The major duties of the DASD (JC) include responsibility for ensuring the communications environment is included in future plans horizontally across DoD as well as coordinating and synchronizing communication efforts in the interagency. This new function and associated developments will start DoD on the path to identify ways and means to fill communication gaps, coordinate and synchronize efforts, and establish a robust and credible communications capability.

The recently published 2006 Quadrennial Defense Review (QDR), on the very last page, finally does address strategic communications and recognizes the need to "improve integration of this vital element of national power into strategies across the Federal Government."[10] The DASD (JC) has been given the responsibility to coordinate DoD integration within the interagency

---

6   Gen. Richard B. Myers, *National Military Strategic Plan for the War on Terrorism* (Washington, D.C.: The Pentagon) p. 32-33.

7   Department of Defense, *National Military Strategic Plan for the War on Terrorism* (U), Annex H, Strategic Communication, (Washington, D.C.: The Pentagon, 18 April 2005).

8   Secretary Donald H. Rumsfeld, "Remarks at Council on Foreign Relations," 17 February 2006; available from http://www.defenselink.mil/transcripts/2006/tr20060217-12538.html.

9   Rear Adm. Frank Thorpe, Deputy Assistant Secretary of Defense (Joint Communications) (DASD (JC)) duty description, e-mail message to author, 10 February 2006.

10   Secretary Donald H. Rumsfeld, Quadrennial Defense Review Report, 6 February 2006 (Washington, D.C.: U.S. Department of Defense, 2006, 92.

community and lead the actions outlined in the recently approved spin-off of the QDR, the Strategic Communication Execution Roadmap.[11] The roadmap is intended to provide strategic direction, a plan of action, and milestones to close capability gaps identified during the QDR development process. In order to close the gaps DoD will organize, train, equip and resource key communication capabilities as described in the roadmap.

Now that strategic communication is being addressed in substantial reports and strategies, synchronization of communication and public diplomacy efforts across the services and among various elements of national power may be dramatically improved. Combatant commands are currently executing theater security cooperation plans without the benefit of synchronized effort with State Department communication objectives. Combatant Commands are now tasked to include strategic communication annexes in crisis and contingency plans, security cooperation activities, and military support to public diplomacy.

Former director for Strategic Communications and Information on the National Security Council said, "All elements of the plan should be designated to help achieve political, economic, and military objectives for the region. Coordination mechanisms include elements of the combatant command staff (operations, intelligence, strategy and plans, public affairs, strategic communications, information operations, Psychological Operations, and Civil Affairs and the Staff Judge Advocate), U.S. Embassies (foreign policy, intelligence, State Department public diplomacy affairs, Defense Attaches, and regionally oriented USAID advisers) and to the extent possible, allied representatives."[12] Ultimately, combatant commanders will be able to tie communication operations and efforts of military support to public diplomacy to an overarching strategic information plan and State Department efforts in public diplomacy.

Vertical and horizontal coordination and synchronization

The recent debate in the political arena and news editorials over the use of paid positive news stories in Iraqi newspapers is a good example of the need for synchronized communication strategies, plans, and policies. The Secretary of Defense and the President's reported concerns about the pay-to-print program led to controversial statements by each of them. Closely

11 Rear Adm. Frank Thorpe, Draft Quadrennial Defense Review Execution Roadmap for Strategic Communication, e-mail message to author 10 February 2006.

12 Jeffrey B. Jones, "Strategic Communication, A mandate for the United States," *Joint Forces Quarterly*, 4th Quarter 2005, 110.

coordinated strategies, plans, and policies would improve understanding and expectations of procedures at every level and help avoid unnecessary public criticism.

Perception of wrongdoing can become reality for some people, as *Los Angeles Times* reporter Mark Mazetti illustrates in his criticism of the military in Iraq for abridging the principles of the 1st amendment to the U.S. Constitution which the military is sworn to defend. "The military's effort to disseminate propaganda in the Iraqi media is taking place even as U.S. officials are pledging to promote democratic principles, political transparency and freedom of speech in a country emerging from decades of dictatorship and corruption."[13] Poorly coordinated information operations failed to account for second and third order effects of a seemingly harmless Psychological Operation initiative. These actions introduce the possibility that the clear distinction between public affairs and information operations will be blurred, and potentially damage American credibility. The effort was intended to help promote goodwill between the Iraqis and the coalition. Instead, it provided opportunity to critics of military operations in Iraq, and ammunition to the insurgent's war of ideas. Secretary Rumsfeld's comments on this controversial issue led to criticism and questions of credibility recently when he told reporters that the pay-for-print program had been stopped, when in fact it had not.[14] With better vertical and horizontal coordination, and an established strategic communication strategy, the DoD and its combatant commands will be empowered to develop more proactive shaping efforts, and avoid the communications impact of disjointed planning.

The recent U.S. air strike in Pakistan that targeted bin Laden's deputy, Ayman al-Zawahri, is a good example of poor interagency coordination. The initial reports of the bombing left both the Pakistani government and the U.S. military in the awkward position of having to explain actions for which neither had been informed nor prepared in advance to respond. Close coordination within the interagency community is imperative to plan and execute a coordinated information campaign that supports GWOT objectives.

Capability gaps

The DoD must reprioritize resources and accelerate implementation

13 Mark Mazzetti and Borzou Daragahi, "U.S. Military Covertly Pays To Run Stories In Iraqi Press," *Los Angeles Times*, 30 November 2005, p. 1.

14 Robert Burns, "Rumsfeld: planting stories under review," Associated Press, 21 February 2006.

of the recommendations of the Strategic Communication Execution Roadmap in order to close identified capability gaps and gain an advantage in the contemporary global information environment. "Operations in Afghanistan and Iraq have underscored that we must create a greater capacity to capture still and video images and develop improved means to transmit, package and use them imaginatively."[15]

Until June 2004, military public affairs units and practitioners did not possess the capability to transmit video images of broadcast-quality in real time. Very small video image clips could be transmitted through the internet, however, larger video image packages had to be mailed through the postal system or delivered by courier. By the time these materials reached their destination they were of no news value. In the early stages of Operation Iraqi Freedom, Third Army Public Affairs purchased suitcase video teleconferencing systems that could be connected to International Maritime Satellite (INMARSAT) telephone systems to transmit video images in near real time.[16] The quality of the video image transmissions was very poor due to the limitation of the systems to transmit only 128 kilobits per second. However, they did prove useful in conducting live interviews with commanding generals in Iraq direct to the Pentagon Press Corps. This experience revealed a communications gap in military public affairs capability.

Over the past eighteen months, the Third Army Public Affairs staff has successfully applied lessons learned from Afghanistan and phase one of Iraqi Freedom to establish a unique, emerging capability to help bridge the military communications gap. This effort is making dramatic impact on its ability to get broadcast-quality video products to the American and coalition publics through commercial news organizations. This Digital Video and Imagery Distribution System (DVIDS) capability has dramatically increased the quantity and quality of news stories about operations in Southwest Asia being told in the local and regional news outlets all across the U.S. and in some foreign countries. Local and regional news agencies that do not have the funding to send correspondents or reporters to the war zone now have the ability to conduct live interviews with soldiers from their home towns who are currently serving in Afghanistan, Iraq, Kuwait or Qatar.

A military distribution center, established at a commercial teleport

---

15 Jones, 112.

16 The author's personal experience as Deputy Public Affairs officer for Third U.S. Army in Operation Iraqi Freedom is the source of background information and developments involving the inception of the Digital Video and Imagery Distribution System.

in Atlanta, serves as the conduit for transmitting broadcast quality video from units in the CENTCOM AOR to local and regional media throughout the U.S. More importantly, this is a two-way system. Small local and regional news networks can conduct live interviews from their studios in the U.S. through the hub in Atlanta to soldiers from their local area serving in Iraq, Afghanistan, Kuwait and Qatar. Any news organization can register for free and download any of the products from this service. News networks can arrange to pull down video files or a live signal by satellite link or through fiber networks free of charge.

The military provides a plethora of fresh news as well as video and still digital images daily from Southwest Asia that media can use freely in packaging their news stories. Or, the media can initiate live interviews through the system to create their own stories. A multitude of stories are released every day and published on military websites for media to take freely, but few are ever used in the national media. On the other hand, local and regional media are hungry for this information and increasingly use this service. The local and regional markets are interested because they can access stories about units, and interviews with soldiers, from their local area that are of great interest to their respective audiences. DVIDS enables Third Army to deliver stories and video products to interested news markets that national media does not have time to tell in daily news cycles, and would otherwise go untold.

DVIDS has already been used numerous times to facilitate White House, DoD and Army public communications requirements from Iraq and Afghanistan, but as yet there has been no effort at those levels to adopt this as a permanent strategic capability. Joint doctrine states that, "The Joint Forces Commander (JFC) must ensure that the Public Affairs (PA) infrastructure to support the joint mission is, to the maximum extent possible, compatible with current media technology."[17] DVIDS enables the military to deliver industry-standard broadcast-quality products to national and international media in real time from anywhere on the globe.

Since DVIDS inception in June 2004 media have downloaded more than 72 thousand video files, conducted more than 1,500 interviews with soldiers in the combat zone, and completed more than 5,400 media requests through this system.[18] These statistics are growing exponentially as more

17 Department of Defense, Public Affairs, Joint Publication 3-61, (Washington, D.C.: U.S. Department of Defense, 9 May 2005) page V-3.

18 LTC William Beckman, Director of the Digital Video and Imagery Distribution System hub, Atlanta, Georgia, e-mail message to author, weekly statistics, 3 Mar 2006.

systems are purchased and distributed throughout the Central Command area of responsibility (AOR).

From the six original systems purchased initially, there are now 52 systems in use in six countries in Southwest Asia, or with units preparing to deploy to the region. The Army owns the predominance of the systems, but recently NAVCENT also purchased a system to use from its headquarters in Bahrain, and Marine units are using DVIDS in their respective sectors in Iraq.

Unfortunately these systems are only being used in the CENTCOM area of responsibility. Initially, this system was intended to serve domestic audiences; however, it has grown in capability to support our coalition partners with links to their home nations. It has also been used during humanitarian relief operations after earthquakes in Iran and Pakistan, as well as the Multi-national Bright Star exercise in Egypt. This system can and should be employed by all services world-wide to support communication objectives during the full spectrum of military operations, from humanitarian relief efforts to high intensity warfare.

This system is currently operated and maintained primarily with Operation Iraqi Freedom funding which could expire instantly if military supplemental funding dries up and the current contract expires. While key public affairs leaders in DoD, JFCOM and CENTCOM strongly advocate the benefits of this capability, little action is currently underway to resource it and incorporate it into the core capabilities of defense public affairs. In order to expand this capability to support global efforts of the joint public affairs community it must be entered into the capabilities acquisition process to be established and resourced as a program of record.

DVIDS must be resourced with personnel to give the DoD 24/7 capability to operate proactively in the global information environment to counter the messages of extremists and terrorist groups in the world media.

The QDR Strategic Communication Roadmap identifies programmatic and budgetary implications of Strategic Communication initiatives and tasks the Army with maintaining DVIDS until the next generation program of record is developed.[19] But it doesn't go so far as to specify who will take on responsibility for initiating actions to enter the capability requirement into the acquisition process. This should be assigned to Joint Forces Command (JFCOM) in order to ensure this capability is expanded to support all combatant commands.

In a parallel action to Third Army's DVIDS, Joint Forces Command

19 Thorpe, Draft QDR Execution Roadmap, p. 2.

has initiated efforts to establish a deployable Joint Public Affairs Support Element (JPASE) to provide rapid response communication capability to augment Joint Warfighters in crisis or contingency operations world-wide.[20] The JPASE has a wing that is deployable within 48 hours that can help Joint Force Commanders interface with the media and the public with a DVIDS transmitter tied into the DVIDS distribution hub. This will be a helpful augmentation package to expand capabilities of the combatant commands' regional public affairs staffs, but the Army already has Public Affairs Detachments (PAD) and Mobile Public Affairs Detachments (MPAD) for this very purpose. If properly equipped for interfacing with foreign and U.S. domestic media, and put on a rotational rapid deployment status, the Army public affairs units could easily fulfill this mission.

The joint composition of the JPASE unit gives it a unique advantage of greater breadth of knowledge for dealing with the various services. However, deploying any of these units half way around the world to support a crisis situation would not be timely enough to meet the immediate needs of real time news reporting. It would be better to provide DVIDS systems to all combatant command organic public affairs staffs to establish an immediate response capability in-place in every forward region that can respond within hours rather than days. Then the JPASE or Army public affairs units could expand their operations within two or three days.

Consolidating and synergizing public affairs assets

In concert with the current round of Base Realignment and Closure plans there is an immediate opportunity to leverage several DoD Public Affairs resources and change the way military public affairs is structured. For example, the Army and Air Force Hometown News organization and the DVIDS capability are currently separate operations performing similar functions in two locations in the U.S. Significant efficiencies in dollars and personnel can be achieved by co-locating and combining the resources of these organizations. Considerable developments in technology provide the opportunity to merge these capabilities and capitalize on technical developments in the private communications industry. The Hometown News organization is still flying video teams to various locations, including Iraq and Afghanistan, to collect video greetings for distribution to U.S. news organizations during holiday periods.

---

20 Rati Bishnoi, "Deployable Joint Public Affairs To Help Commanders' Media Outreach, (Washington, D.C.) *Inside The Pentagon*, 9 February 2006, p. 1.

There are already public affairs units in these and many other countries that can perform this function and transmit the video products via their DVIDS satellite transmitters to a central hub in the U.S.

In just one holiday period, the DVIDS hub received and processed 6000 greetings in comparison to 300-400 acquired and processed by the Hometown News teams in the same period. The Army alone could achieve significant cost savings in resources and personnel by combining the expertise of the Hometown News video processes with the capabilities of the DVIDS hub. The Navy, Air Force and Marines could also reap tremendous benefit from joining this venture. DVIDS can serve all combatant commands, all services, and DoD by expediting the flow of accurate and timely information about the activities of U.S. joint forces to the public and internal audiences.

Additionally, the Army is working on creating a Soldiers Media Center (SMC) at Ft. Meade, Maryland to consolidate several PA assets from the Washington, D.C. area, improve synergy, and reduce overhead costs. DVIDS can serve as the backbone for SMC by providing established infrastructure and technical capability through the Crawford Communications teleport. Making DVIDS a Program of Record (POR) through the joint acquisition process will ensure continued funding for maintaining today's Visual Information (VI) capability and developing tomorrow's VI capability.

Leveraging a commercial teleport will avoid the expense of building a new military facility, and purchasing satellite systems and other equipment needed to receive and distribute video and print materials. The tremendous benefits of outsourcing this service and leveraging the spending power and technical investments of a private company could not be matched with the expense of military construction, purchasing equipment, and training service members to perform this mission. In the long term, DVIDS should be made a POR and all services should be encouraged to combine efforts to join and expand this venture to improve synchronization and integration of this vital capability in all plans and operations.

Non-traditional audiences and news venues

New public audiences are emerging through communications systems involving the internet or electronic messaging, creating new news audiences in non-traditional news venues. The future leaders and decision makers of the world include a growing number of people who have grown up playing video games and clicking through multiple media presentations. The developing audiences in foreign nations have also found the internet and emerging technologies to be a window to news previously not available to them. In order to grab their attention it is necessary to develop ways to present news that engages and involves them.

Defense Secretary Donald H. Rumsfeld said, "Insurgents around the world have learned to use this media-rich environment to further their goals, and to successfully combat them, the U.S. government needs to adapt and use the technology that has proven to be so powerful...." He goes further to say, "forces deployed to a theater of operation need to be experienced enough to engage the full range of media that the world uses." This will require an elevation in Internet operations, the establishment of 24-hour press operations centers, and training in other channels of communication," he added.

News groups and listserves are providing the information exchange for which people went to the commercial news organizations in the past. Access to foreign press sites, and alternative press are letting news foragers go beyond the local or national news franchise perspective on the news. The conventional model of a mass medium was a one-way flow of the news from the news organizations to the public audience. Things like weblogs are providing a different way to provide news selection and commentary. Bloggers invite the contributions of their readers, and facilitate conversation between the readers. The discussion generated about the news item is often of the most interest, and provides the best understanding of the news event, rather than the news item itself. "Citizens are interested in participating and contributing to subjects that traditional news outlets ignore or do not often cover. Clyde Bentley, an associate professor at the Missouri School of Journalism, notes, "The main difference between traditional journalism and citizen journalism is that traditional journalists are sent out to cover things they don't really care about; in other words, the next city council meeting isn't going to make or break their lives. But a citizen journalist is not out to cover something, but to share it. For them, they want to tell everybody about their passion."[21]

Independent newspapers, magazines, websites, radio and TV are becoming more widely available and emerging news sources are building their own audiences worldwide. In the contemporary automated information environments people increasingly receive news and information through web sites, electronic mail, web logs (blogosphere), and personal digital assistants

---

21   Shayne Bowman and Chris Willis, "Nieman Reports: The Future Is Here, But Do News Media Companies See It?" Nieman Reports, Volume 59 Number 4, Winter 2005, a publication by The Nieman Foundation for Journalism at Harvard University.

(podcast).

Podcasts are an easy way to download available remarks from government officials, daily press briefings and listen to them on a personal media player. Apple computers launched the first capability for consumers to download music, news and other entertainment to iPods for users to take with them on the go. Since the creation of this capability its menu has grown to include 22 primary categories and 56 sub categories, however, there is currently no military category among the iTunes options. The DoD could miss the opportunity to gain an edge in this emerging capability if it delays in establishing its position as the preeminent provider of military materials in this new media. This leaves great potential for another break in the communication gap identified in the QDR if the department is not proactive in capitalizing on this market.

Recently Google has announced its initiative to enter into this market and therefore create another potential audience for military patrons. Google Inc. is upping the ante in the online video gold rush, allowing content owners to set their own prices in a bid to create a more flexible alternative to Apple Computer Inc.'s pioneering iTunes store. Google's planned video expansion, announced Friday at the Consumer Electronics Show, already has lined up commitments to sell thousands of downloads, including recent television broadcasts of popular CBS shows and professional basketball games, as well as vintage episodes from series that went off the air decades ago. A launch date for the expansion has not been released.[22]

As technology rapidly emerges, so does the potential for new nontraditional news markets that may dominate conventional systems as the next generation enters the working world and gains interest in the greater world. Really Simple Syndication (RSS) is another easy way to get the news whenever it is updated even if not logged on to an internet news source. With this system, various governmental organizations provide continually updated headline feeds and deliver them directly to one's desktop news reader. One can install a news reader that displays RSS feeds from personally selected Web sites. Like an email program or a Web browser, the news reader serves as a kind of information portal, and provides a real-time interface to the selected feeds. Once the RSS reader is set up to access a news source, it continuously checks the source's RSS feeds for the latest headlines. A number of free and commercial news readers are available for download from the internet.

List Serve is another variation of automated news access. One can automatically receive via e-mail full texts of selected government documents

---

22 The Associated Press, "Google opens doors of online video bazaar," 7 January 2006.

and publications that provide key official information. In order to keep pace with communicating to these emerging audiences it is critical to adapt communication strategies and technologies to participate in these environments. It is also necessary to be able to accelerate responses to adversarial misinformation and disinformation to all potential audiences. Military public affairs programs have made piecemeal jabs into these technologies but have not made a broad effort to engage aggressively in these new areas aside from the plethora of websites for every unit and organization. Central Command public affairs has established a deliberate effort to engage in the blogosphere as an official organization, while some others have responded individually under official titles to correct misinformation in various web log forums.[23] DoD has done some analysis of trends in many of these internet based technologies, but has yet to enter formally into information dissemination by these nontraditional means.

Foreign audiences

In the blurred transition to phase four of OIF the Organization for Reconstruction and Humanitarian Assistance (ORHA) was expected to take on the responsibility for communications with the local population. Military public affairs units worked to fill the gaps in various provinces to varying degrees of success, but ORHA did not begin actual communications to the local population for a full two months after arrival in Baghdad. In the meantime, neighboring countries began broadcasting and publishing conflicting information and agendas to fill this void. The military should be expected and equipped to fulfill communications needs for the safety and security of the local population until the security situation enables other systems to be emplaced or restored.

Public affairs plans and capabilities must include preparations and means to provide information to the local populace in foreign areas of operations at the earliest opportunity in warfighting or peacekeeping operations. This will require translators who are already embedded with the public affairs staff prior to initiation of hostilities. It will also require special equipment to set up printing operations, radio transmissions, and television production in the formats and language of the local population.

Translators are in short supply and competing demands for their critical services in intelligence, civil military operations, and other requirements

---

23 Capt. Steve Alvarez, "CENTCOM Team Engages 'Bloggers'," American Forces Information Service, 3 March 2006.

in the battle space quickly absorb all available linguists. It must be recognized that in the global information environment, and battle of ideas, public affairs and information operations must have early allocation of linguist support. This requirement is not just for translation, but for understanding the culture and nuances of the indigenous population as themes and messages are initially developed. It is not sufficient for the linguist just to be able to translate the language for written and broadcast materials. It is critical that the linguist be from the region of operations so they can properly craft and translate messages that will resonate with the target audience without offending or alienating them.

Printing requirements include computers with software and/or keyboards that enable translation into the native language. Ample printing materials must be brought in at the earliest possibility to initiate and sustain printing requirements until services and supplies can be purchased on contract through the local civilian market and normal production can be resumed by new or existing commercial publications.

Similarly, radio and television broadcast capabilities must be planned and ready to implement immediately to inform local populations where to get assistance, encourage them to comply with rule of law, and cooperate with coalition military efforts to restore basic services and return control to civil authority. Part of the challenge for ORHA was moving about the country and setting up communication stations in a tenuous security environment in which ORHA was unable to provide its own security.

A team from the 22nd Mobile Public Affairs Detachment (MPAD), supporting the 101st Airborne Division in Mosul, worked with the civilian operator to restore the radio station and support the commander's communication with the surrounding population. This proved beneficial to the early success achieved in that region. However, Baghdad was a different story. U.S. efforts to stop Iraqi propaganda in the initial stages of combat operations collapsed the government-controlled public information system. The government communications facility was completely destroyed, and ORHA's replacement equipment would not arrive and be established for more than two months. This communication gap opened the door for opposition communications to undermine the coalition and interim authority.

Secretary Rumsfeld recently said, "Let there be no doubt, the longer it takes to put a strategic communication framework into place, the more we can be certain that the vacuum will be filled by the enemy and by news informers that most assuredly will not paint an accurate picture of what is actually taking place."[24] The current Table of Organization and Equipment for deployable public affairs units is not adequate to meet the capability requirements for interacting with foreign audiences. It is evident from operations in Iraq and Afghanistan that military PA units must possess the ability, equipment and translators, to inform indigenous audiences in their respective areas of operations. The lessons learned and experiences of PA assets in active theaters of operations, most of which are Army units, must drive doctrine, organization, training, materiel, logistics, personnel and facilities (DOTMLPF) planning. The Army Public Affairs Center should review the mission and equipment for deployable public affairs units and initiate actions to make appropriate changes in DOTMLPF as necessary to fill this gap.

While there are signs of progress, the DoD efforts in strategic communication currently lack adequate strategic direction, and interagency coordination. The accomplishment of military objectives is at great risk, as well as the potential for squandering national resources in costly military campaigns without a coherent communications strategy. There must be a top-down strategy from the President to synchronize efforts of diplomacy and public information with activities of various other elements of national power to support foreign policy objectives. The information campaign must be a continuous process in peacetime, during military campaigns, and throughout stability or peacekeeping operations.

"We are fighting a battle where the survival of our free way of life is at stake and the center of gravity of that struggle is not simply on the battlefield overseas; it's a test of wills, and it will be won or lost with our publics, and with the publics of other nations," Secretary Rumsfeld said. "We'll need to do all we can to attract supporters to our efforts and to correct the lies that are being told, which so damage our country, and which are repeated and repeated and repeated."[25]

The urgency of the need for a proactive and responsive communications strategy warrants accelerated DoD actions to synchronize efforts, engage in non-traditional media, and develop creative solutions for improving strategic communications. In combination with making the DVIDS service a DoD-wide capability, there is an immediate opportunity to leverage several Public Affairs resources and change the way military public affairs is structured to gain synergy from combining the various capabilities. Strategic communications is

24 Rumsfeld, Remarks at Council on Foreign Relations, 17 February 2006.

25 Ibid.

a critical enabler for the employment of the national elements of power to achieve political objectives. It is past due time to transform DoD strategic communication in order to support military objectives effectively through communications strategies. Diplomatic, economic, and military actions reinforced by strategic communication are necessary to advance national policy objectives: "No single contributor is preeminent. All are required in a synchronized and coherent manner."[26]

---

26 Jones, 114.

17

Islamism and Stratagem

John J. Dziak

Introduction

This essay is intended as a running reflection on the intelligence and counterintelligence problems posed by resurgent, militant Islam, especially in the post-9/11 era. It is not meant to be a comprehensive assessment either of Islam itself or a prescription for the Intelligence Community on how to portray militant Islam as an intelligence issue for policy makers. Rather, it offers observations on these intelligence problems with an effort to identify and make sense of the deceptive dimensions therein. Such dimensions include not only active deception and manipulation activities undertaken by our adversaries, but the proclivities inherent in the doctrinal aspects of militant Islam itself. In that sense Islamic stratagems against the West are both explicit and implied within the theocratic culture of Islam. But they also embrace a tendency among much of the media and within U.S. policy and intelligence to deny that there's even a problem in the first place.[1] These reflections draw on the author's experience in intelligence and counterintelligence and graduate school teaching of the same subjects for over four decades. Obviously, the views are those of the author alone and are meant as a stimulant to a discussion of timely and contentious intelligence

---

1 Much of the mainstream media tend to refrain from any Islamic adjectival association with terrorism, in some cases even eschewing the use of the term "terrorist" in favor of the more benign "insurgent". Pronouncements from the White House likewise, for several years after 9/11, decoupled "Islamic" from "terrorism" in discussions of the war we are in. This desire not to explicitly name culprits and thereby not offend even extends to non-Muslim actors. For example, the recent (March 2006) mild reaction of the U.S. government to the released Iraqi intelligence reports of alleged Russian penetration of CENTCOM and the passing of U.S. warplans to Saddam Hussein is suggestive of such an attitude. The object here apparently was not to antagonize President Putin and his government.

issues related to hostile deception.

Militant Islam (also frequently identified as Islamism, radical Islam, fundamentalist Islam, Islamo-fascism, etc.) arguably is the pressing U.S. strategic problem of the moment and promises to remain so for an indefinite period.[2] Ever since the 9/11 attacks, however, major elements of the U.S. leadership, policy, media, and academic elites have danced around the issue of clearly identifying or naming the enemy. Agreement on a generally accepted terminology for the present war itself is problematic. By attempting to define down a national security problem so as not to offend the larger faith community of Islam, or implicate it in the spreading terror attacks worldwide, opinion molders tend to seek refuge in highly inaccurate and misleading euphemisms such as the generic "terrorism" or "terrorists." But terrorism is a technique, not an enemy. If, for instance, a similar plague of political correctness were in vogue in World War II, we would have been fighting the "perpetrators of blitzkrieg who needed to be brought to justice" or the "practitioners of surprise attack," rather than the German Nazis or Imperial Japanese militarists.

A number of highly visible bloody actions in 2004-2006 in which Islamic driven anti-Western hysteria was even further whipped up, hopefully may have given pause to those exponents of such timorous Western behavior: to wit, the brutal murder of Dutch filmmaker Theo Van Gogh, the 2004 train bombings in Madrid, the London bombings of summer 2005, the fall 2005 riots in France, the highly organized Danish cartoon frenzy in early 2006, and the case of the Christian convert Abdul Rahman in Afghanistan in March 2006. Equally disturbing to Western sensibilities was the favorable resonance these events generated among large numbers of Muslims throughout the world, especially among the immigrant Muslim populations (significantly, among the second and third generations) in England and elsewhere in Europe. What these events should force us to confront is that Islamic doctrine as found in the holy or revered texts (Koran, Hadith, Sira, etc.) of that faith is at variance with the core tenets of Western civilization and its survival. As aptly put by one observer, "... the Shariah principles in question are shared by all four of the Sunni schools of jurisprudence (Maliki, Hanbali, Hanafi, and Shafi'i), plus the Shia school. There is no 'sixth' school that recognizes religious and civic freedom, in any way what these expressions mean in the West."

All five of the actual schools or traditions take a view of idolatry

---

2 This is not to diminish other pressing strategic problems such as an increasingly assertive and militarily powerful People's Republic of China. But the PRC is not the culprit in 9/11 and its aftermath.

that entirely removes the possibility of freedom of expression in public life. Moreover, all take a view of apostasy that presents a palpable threat to the life and liberty of every non-Muslim, and excommunicated Muslim. And such doctrines as "jihad" (when interpreted as holy war against all infidels), and "razzia" (permission to raid and plunder our infidel communities) are not such as can be assimilated with Western jurisprudence.[3]

That may be one of the many reasons why it is so difficult for so-called Muslim moderates to organize to soften or reform the scriptures of that faith. Aside from theological proscriptions forbidding changes to the unmediated word of Allah as revealed to the Prophet, there is no central institutional authority within Islam to facilitate or foster such changes even if a broad desire were present. It is unfortunate that such candid recognition of Islamic theocratic realities are not part of a broader public and official dialogue in the West in general and in the U.S. in particular, since its absence also affects the ability of the Intelligence Community itself to come to grips candidly with the strategic implications of a resurgent, imperial, and militant Islam.

The inability or refusal of a putative victim to admit that he even is a target of another's hostile attentions and attendant deception is by no means restricted to Western elites. The arch-deceiver of the Twentieth Century, Joseph Stalin, so valued his non-aggression pact with Hitler that, for all practical purposes, he deceived himself into absorbing Hitler's design for Moscow's defeat well before the June 1941 invasion. A recent compelling account of the reasons for the initial success of Operation Barbarossa put it this way: "… it was Stalin's insistence on accepting German deception as truth, his rejection of valid intelligence from his own services, and his failure to recognize that the warnings from Western powers, themselves threatened by Hitler's aggressiveness, were both accurate and well intentioned, that led to the debacle of the summer of 1941."[4]

In what may be an apocryphal account, Lenin, in the early 1920s, is said to have advised his chief of counterintelligence, Artur Artuzov, to tell the Whites and the West what they wanted to hear when it came to devising the "legend" for the CHEKA's classic strategic deception operation, the "Trust." Even if the above tale is itself "legend," Stalin should have remembered

---

3 David Warren, "Staring Down Shariah", Real Clear Politics, March 30, 2006, http://www.realclearpolitics.com/articles/2006-03/staring_down_shariah.html.

4 David E. Murphy, *What Stalin Knew: The Enigma of Barbarossa*, New Haven: Yale University Press, 2005, p. 249.

the actual "Trust" operation, for he too had a major role in its conception and conduct, and would have understood well the credulity of his Western opponents and their susceptibility to being gulled. The Whites and the West self-deceived in their response to Artuzov's "Trust" deception of the 1920s; Stalin self-deceived before Hitler played him for the fool in 1941. The West's failure with the "Trust" and Stalin's failure with Barbarossa are lessons worth revisiting by Western leaders faced with a resurgent Islam – an ideology with almost fourteen hundred years of experience in empire, stratagem, and dissimulation, not unlike the totalitarian "isms" of the Twentieth Century.

In view of the above experiences it is useful to reflect that an intimate connection between counterintelligence and deception has always existed, especially in non-Western systems. Whether in traditional Islamic societies, autocratic or totalitarian ones like Tsarist and Soviet Russia, or despotic systems like the ancient hydraulic societies of China or the contemporary People's Republic of China, a counterintelligence system was the intelligence service of choice.[5] Counterintelligence was and is essential to the maintenance of autocratic, totalitarian and despotic elites whose singular claims to monopoly rule could be upheld only through secret police-type operations – hence "the counterintelligence state" in which a fixation with enemies and threats to the power of these elites was the reigning ethos. Penetration, provocation, deception and other active measures characterized the operational tradition of the "counterintelligence state." This tradition carried over to methods of warfare as epitomized, for instance, by the Mongol conquests of Eurasia in which enemy attention was masterfully deflected by ruse, false threats, and myriad stratagems resulting in stunning victories, frequently against superior forces. In such traditions deception is not merely a military plan annex, dusted off for a particular military campaign, but rather is organic to the whole culture of a counterintelligence focused system. It is part of the DNA of traditional societies and cultures, including Islamic ones. Whether one looks at the religious police of Saudi Arabia, the secret police services of Syria or Iran, or the operations of Al-Qaeda one sees the hallmarks of a counterintelligence system, not just Western style intelligence services.

---

5   For an in-depth treatment of traditional, despotic systems see Karl A. Wittfogel, *Oriental Despotism: A Comparative Study of Total Power*, New Haven: Yale University Press, 1957. For the linkage between these systems and more recent examples of the "counterintelligence state" see John J. Dziak, *Chekisty: A History of the KGB*, Lexington, MA: Lexington Books, 1988.

The setting

There are numerous facets of Islam's almost fourteen centuries of imperial experience that could be linked to its practices of stratagem and the West's response thereto. Two of these are highlighted as of special relevance to this subject: Jihad and Dhimmitude. The universal goal of Islam as handed down by Allah's prophet Muhammad, was a global order in which all men recognized the rule of Allah as exemplified in Muslim rule, either as believers or as inferior subjects, i.e., Dhimmis. This was to be accomplished through "struggle [or striving] in the way of Allah", later to become known as Jihad. Ibn Khaldun, the noted 14th-century Islamic jurist and historian put it thus: "In the Muslim community, the holy war [jihad] is a religious duty because of the universalism of the [Islamic] mission and the obligation [to convert] everybody to Islam either by persuasion or by force.... Islam is under obligation to gain power over other nations."[6]

When attempting to mask the war context of "struggle in the way of Allah," apologists for Islamic militancy insist that a personal interior struggle for goodness and self-improvement is what is meant. Islamist advocates use the same line on western audiences but clearly mean holy war when referring to Jihad to their Arabic listeners. And this tactic seems to be the most common deceptive technique in the public and diplomatic realm. But both the history of Islam and the justifications of the Koran and the Hadith leave no doubt that the doctrine of Jihad means that the relations between Muslims and non-Muslims are defined by war between two irreconcilables, the dar al-Islam (the land of Islam) and the dar al-harb (the land of war), or the non-Muslims. Jihad constitutes the Islamic ideology of peace and war. Universal peace in an absolute sense as viewed in the West is a non sequitur; only submission to Islamic domination produces a true stable peace. Peace, as viewed from London, Washington, or Berlin falls for Islam into the realm of passing and temporary truces, invariably encrusted in stratagem and dissimulation.[7] While Western commentators fixate on "peace," Islam requires submission (which is what it means) – to Allah and, hence, to his earthly viceroys.

Dhimmitude, or the Islamic system of governing unconverted non-Muslim populations conquered by Jihad wars, entailed existence in a clearly

---

6   Ibn Khaldun, *The Muqudimmah: An Introduction to History*, trans. by Franz Rosenthal, Vol. 1, NY: Pantheon, 1958, p. 473.

7   For deeper treatment of Jihad in history, see: Andrew Bostom (ed.), *Legacy of Jihad: Islamic Holy war and the Fate of Non-Muslims*, Amherst, NY: Prometheus Books, p.28; and Efraim Karsh, "Islam's Imperial Dreams", *Commentary*, April 2006, pp. 37-41.

subservient social status and political disenfranchisement including slavery, under predatory and humiliating taxes (e.g., jizya) and other forms of overt discrimination against so-called infidels. Over the centuries the severity of the system ranged between fairly benign patronizing, to the kidnapping of children and genocidal pogroms, and its net effect was to foster a Muslim arrogance to its non-Muslim minorities, which carries over to relations with non-Muslim nations. On the receiving side of Dhimmitude, an unease and defensiveness seemed to characterize Christian, Jewish and other minorities forced to live a tenuous existence if they elected not to convert to Islam. Dhimmi behavior ranged from, among others, servile submission, to elaborate deceptive practices to survive, to defensive emigration as in the cases of the Jewish Diasporas after the various Arab-Israeli Wars or as in the case of Christian emigration from the Palestinian areas and Lebanon – and now Iraq. It also has been noted by the principal scholar of Dhimmitude, Bat Ye'or, that a servile, fearful, and self-deceiving Dhimmi mindset seems to have been absorbed by some Western political and social elites in their dealings with contemporary Islamic minorities, groups, and nations in response to assertive and threatening Islamic behavior, whether domestic or international. The craven and hypocritical response to the Danish cartoon controversy by large segments of the media and academia as well as Western governments – the putative guardians of free speech – is a troubling indicator of just how far the servile psychology of Dhimmitude has advanced in the non-Muslim world. This has not gone unnoted by Islamic militants as witnessed by the adroit way they used the Madrid bombings to help bring down the Spanish government. Jihad and Dhimmitude are forged together in history.[8]

Sources of contemporary Islamism – internal

The sources of contemporary Islamism are worth a brief examination before viewing the practice of stratagem in the Islamic tradition. What should come as no surprise is that militant Islam, or fundamentalism (itself a Western projection from its own religious experience), is not a recent product ushered in by the Muslim Brotherhood, the Salafists, or the Wahhabists of the Arabian Peninsula and Al-Qaeda. As already seen with regard to Jihad, Islamist militancy is inherent to Islam itself throughout its almost fourteen hundred year history. What the above movements, and states like Shi'ite Iran, are now

---

8 See for instance: Bat Ye'or, *Islam and Dhimmitude: Where Civilizations Collide*, Madison, NJ: Fairleigh Dickenson University Press, 2002; and Bat Ye'or, *Eurabia: the Euro-Arab Axis*, Madison, NJ: Fairleigh Dickenson University Press, 2005.

advocating, is a return to the purity of the earliest periods of Islamic history to include Mohammed's time, and other "good" eras such as the much touted Islamic golden age of the Middle Ages. Bursts of Islamic "puritanism" occurred throughout history including the long eras of its imperial success – as well as in its period of decline as exemplified by the later years of the Ottoman Empire. This may not be welcome news to Western secular elites whose universal frame of reference is Enlightenment rationalism. But it certainly should be part of the intellectual kit bag of intelligence and counterintelligence professionals.

Further, Islam, as a self-identified theocratic system, may be regarded as a political ideology rather than simply another confession common to the West in which separation of church and state is part of the creedal faith. It is an established system for the integrated governance and faith observance of all peoples according to the Koran. Ideally, the Koran must be learned by heart (which many believers do even if they cannot read or comprehend Arabic, the preferred tongue for doing so) and applied in a literal way. This is because according to Islam the Koran, directly revealed by Allah through the archangel Gabriel, contains all that is needed for this life, and therefore no interpretation a la Biblical exegesis is allowed or tolerated. If Islamic liberalizers or moderates are too vocal in an effort to "reform," or "update," or find a better application of Islam to the contemporary world they run very serious risks of being labeled apostates, thereby risking death. Moderates are doubly conflicted here: tinkering with the holy text is tantamount to questioning Allah's given word thus risking charges of apostasy; to be an apostate invites retribution. Moderates throughout the Muslim world face not only the retributions of Islamists but from their governments as well. The majority of Muslims live in countries ruled by very oppressive regimes; even relatively moderate countries offer little protection from rigorous enforcement of Islamic law. Jordan, for instance, jailed editors for daring to reprint the Danish cartoons. Violence, therefore, is not only politically permissible but divinely decreed per the holy texts. A straightforward reading of the Koran and the Hadith (the traditions and sayings of the prophet) reveals that violence is condoned and advocated, and has been acted upon throughout Islam's almost fourteen centuries of history. Today's Jihadists certainly are no aberration from that tradition.

Additionally, Islam, like traditional Judaism, is an "orthoprax" (correct practice) faith placing "fundamental emphasis on law and regulation of community life." Scholars in general view Christianity as an "orthodox" (correct opinion) faith wherein "greater emphasis on belief and its intellectual structuring of creeds, catechisms, and theologies," is placed.[9] This is not an insignificant difference. In Islam religious debate tends to center on consistency of practice with fixed law, resulting in concentrated focus on Islamic law, Sharia. Legal interpretations are based on precedent that looks to the past and does not presume to probe the nature of Allah or his designs. Among Christians debate tends to focus on doctrine, hence the existence of clergy, hierarchies, and theologians to explore, discover, develop, debate, and explicate doctrine and liturgy – all in a presumptive urge to grasp the nature of God better through the operation of reason.[10] In Islam, the divine transmission of the religious text directly to God's final prophet, Mohammed, fosters literalism, "the Scripture whereof there is no doubt."[11] The Koran is the unmediated word of Allah; the Hebrew Bible and the New Testament are "inspired" texts, mediated through human agency and subject to interpretative discovery. When Jihadists invoke the sacred texts to justify violence either against fellow Muslims or the infidel, they can claim to be in a legally superior position to a moderate Muslim who must strain to adduce a countervailing argument based on the same texts. The latter is at a clear disadvantage.

Sources of contemporary Islamism – external

Looking outside Islam for other sources of influence we must examine the attraction for Islamists of the premier totalitarian movements of the twentieth century, namely Nazism and Soviet Communism. To begin with the former, Islamism has been characterized as fascism with an Islamic face. Islamism and Nazism/fascism have collaborative roots going back to the early twentieth century in mutually perceived and shared practices, grievances, common enemies, and formative catastrophic experiences.[12] Both shared a deep hatred of Christianity, Western culture, capitalism, liberalism, and Jews and America in particular. Although militant Islam long predates Nazi Germany, its twentieth century resurgence paralleled the rise of Nazism in mutually perceived catastrophes. These were the collapse of the Ottoman Empire

---

9 Rodney Stark, *The Victory of Reason: How Christianity Led to Freedom, Capitalism, and Western Success*. NY: Random House, 2005, p. 8. Although it isn't intended as such, Stark's work is an excellent starting point for contrasting the bases of Islam and Christianity.

10 Ibid., p. 9.

11 Ibid., quoting Pickthall's translation of the Koran.

12 This section draws on the following series by Marc Erikson: "Islam, Fascism and Terrorism," *Asia Times*, Parts 1 – 4, 5 & 8 November 2002, 4 & 5 December 2005, http://www.atimes.com/atimes/printN.html.

— the last Caliphate — and the defeat of Germany in World War I followed by the Versailles Treaty.[13] The contemporary variant of Islamism was annealed in the Muslim Brotherhood (Al Ikhwan Al Muslimun) that emerged in 1928 as a direct reaction to the elimination of the Caliphate by the secularist Young Turk reformer, Kemal Ataturk. Founded by Hassan al-Banna and several followers to focus at first on Muslim spiritual reform, the Brotherhood blossomed in the 1930s and 1940s after pursuing far more active political goals and imitating organizational models from Nazi Germany and Fascist Italy. By the end of World War II the Brotherhood had a half million members just in Egypt, not counting the rest of the Middle East. It had modeled itself on Mussolini's Blackshirts with all the paramilitary forces, intelligence elements, and secret apparatus common to both Germany and Italy of that era. And in the struggle for political power it fostered terrorism and political assassinations to the point that the Egyptian government had al-Banna himself assassinated in 1949.

The Brotherhood collaborated with the Germans before and after World II and with another group of Nazi-fascist imitators, the "Young Egypt" (Misr al-Fatah) movement, two of whose notables, Gamal Abdel Nasser and Anwar El-Sadat, became Presidents of Egypt. Like their Brotherhood friends, Nasser and Sadat's "Free Officers" were in contact with German military intelligence before and during World War II, resulting in Sadat's arrest by the British in 1942. An equally important Muslim Brotherhood figure was Sayyid Qutb, frequently billed as the father of contemporary Islamism who helped inspire the likes of Osama bin Laden and his deputy, Ayman al-Zawahiri. Zawahiri is viewed as Qutb's intellectual heir. Qutb wrote numerous radical tracts that became almost canonical readings for Islamists down to the present, including a thirty-volume commentary on the Koran, and the revered Islamist masterpiece "Milestones" in which he propounded an Islamist seizure of the state by an elite vanguard that would then impose Islam from above. Qutb's Bolshevik-Nazi style apparently was radically honed during several years (1948-1951) of graduate study in America whose decadence and female liberties disgusted his Islamist sensibilities. Following release after years in Nasser's prisons, another Brotherhood assassination attempt on the Egyptian leader in the spirit of Qutb's top-down revolutionary recipe, Qutb was rearrested and executed in

---

13 Obviously the roots of Nazism and fascism long predate Germany's World War I defeat. Likewise, the notion of a revived pan-Islamic movement began in the late nineteenth and early twentieth centuries. But Germany's and Islam's mutually perceived losses in the wake of the war were the proximate events which precipitated Nazi and Islamist stirrings and helped to foster a shared identity politics of grievances between them.

1966. Belated revenge, of sorts, came with the 1981 assassination of President Sadat by the Egyptian Islamic Jihad, an offshoot of the Muslim Brotherhood. Together, al-Banna and Qutb, both heavily influenced by the Nazi-fascist model, put their own unique militant stamp on contemporary radical Islam. For years Islamist dissemblers have tried to hide or play down that troublesome pedigree for obvious reasons.[14]

Another key Islamic figure in the Nazi-Brotherhood connection was the go-between for al-Banna and the Nazis: Haj Amin al-Husseini, one-time Grand Mufti of Jerusalem and regarded by many in the Arab world as the founding father of the Palestinian movement and inspiration for the Arab League.[15] Al-Husseini was instigating pogroms against the Jews in Palestine as early as the 1920s, was put on the Nazi payroll in 1936 following a meeting with Adolf Eichmann and, using Nazi supplied funds and weapons, helped initiate the 1936-1939 Arab Revolt in Palestine against the Jews and the British authorities — as well as moderate Arabs and Husseini's Arab opponents. In 1941 al-Husseini had a major role in the failed pro-Nazi coup in Iraq, followed by a murder campaign against Iraqi Jews. Escaping Iraq al-Husseini fled to Germany where he spent the war years hosted by Hitler and treated as a head of state. During his sojourn in Germany al-Husseini campaigned against the exchange of European Jews for German POW's and helped in the recruitment and training of Yugoslav and other Muslims for German sponsored and led military units, including those of the SS.

According to recent research into German wartime records the Germans had stood up a mobile SS unit ("Einsatzgruppe Egypt") in Greece for deployment to Palestine to eliminate the 500,000 mostly European refugee Jews there. Al-Husseini and his Arab supporters were to have an important role in this operation that would have been modeled on the Einsatzgruppe units on the Eastern Front.[16] Montgomery's victory over Rommel in 1942 prevented this from happening. Al-Husseini was heavily involved in atrocities against Jews, Serbs and Gypsies during the war and he was actively sought by

14 For a detailed examination of the Egyptian roots of contemporary Islamism and the roles of al-Banna and Qutb, see: J. Bowyer Bell, *Murders on the Nile: The World Trade Center and Global Terror*, San Francisco: Encounter Books, 2002.

15 For a fuller discussion of the Nazi connection to Islamism, see Matthias Kuntzel, "Islamic Antisemitism and its Nazi Roots", April 2003 (presented at a conference on "Genocide and Terrorism — Probing the Mind of the Perpetrator,"Yale University, 11 April 2003.

16 See *Washington Times*, 13 April 2006; *Boston Globe*, 7 April 2006. The research is found in a new work, published in Germany, titled: *Germans, Jews, Genocide — The Holocaust as History and Present* by Klaus-Michael Mallmann and Martin Cueppers.

Yugoslavia and Britain as a war criminal. After the war the French held him in custody but refused to extradite him. Al-Husseini "escaped" to Egypt with the assistance of the Muslim Brotherhood and spent the rest of his life working against Israel and the West (he died in Beirut in 1974).[17] During that time he and the Muslim Brotherhood worked with German ex-military and security personnel brought in as advisors by King Farouk – to the latter's regret. The Germans conspired with the Brotherhood, Nasser, and his Free Officers to overthrow the King in a well-executed coup, eventually bringing Nasser to power.

One last note on Nazi influence on contemporary Middle East and Islamist developments: the Ba'athist movements in both Syria and Iraq owe a great deal to the Nazi influences vectored into the region by al-Husseini and the Brotherhood. Saddam Hussein in Iraq and the Assad family in Syria were the beneficiaries of that influence, a legacy that both Ba'athists and Islamists would rather be forgotten using the time honored and religiously sanctioned techniques of "taqiyya" (dissimulation or deception) and "kitman" (akin to mental reservation) – more on this to follow. However, Yasser Arafat in a fit of candor a couple of years before his death haughtily dispensed with "taqiyya" when he paid worshipful tribute to al-Husseini as "... our hero ... and I was one of his troops [in the 1948 war]."[18] Arafat's mother was a cousin of al-Husseini, and Arafat supposedly spent four years as a youngster with the Mufti after his mother died. Arafat later became active in the Muslim Brotherhood and the Mufti's own group as well and is rumored to have been involved in running Nazi-supplied arms into Gaza. Arafat's pedigree suggests that it would be wise for western observers not to make too much of a distinction between an Arab nationalism of an earlier generation and a resurgent Islamism of more recent vintage.

When overt Nazi-fascist influence in the Islamist world ebbed after Germany's defeat, Soviet and Warsaw Pact penetration and presence rushed in. Moscow, in a volte-face, switched its support from the Israeli state to the Arabs, reinforcing Nasser's pan-Arab and nationalist schemes (Stalin had, at first, backed the creation of the Israeli state and provided arms). We cannot detail the specifics of Soviet military and political support which lasted through the collapse of Communism; but we will briefly explore the influence of

---

17 A penetrating and concise evaluation of al-Husseini's virulent anti-Semitism and Nazi involvement may be found in David G. Dalin, "Hitler's Mufti," *First Things*, August/September, 2005, pp. 14-16.

18 Al Quds, 2 August 2002.

Soviet/Warsaw Pact intelligence services and their contributions to an already established deception style inherent in Islamist traditions as exemplified by "taqiyya" and "kitman."[19]

The Soviet intelligence and security services were always at the leading edge of any Soviet penetration or aid effort in the Third World, an operational style which simply replicated the way Moscow injected its presence and interests into Eastern Europe at the end of World War II or, indeed, the way it carried out clandestine efforts to spread its revolutionary influence around the world following the 1917 Bolshevik Revolution. The "Organs," as they were known in Soviet parlance, were the Party's instrument of choice in the advancement of Soviet objectives. This continued to the last days of the USSR's existence and carried over into the practices of the current Russian Federation under Putin, the KGB and GRU being followed by the FSB/SVR – and the unchanged GRU.

Wherever the "Organs" went they exported an operational tradition that was based on the nature of Soviet intelligence. The Soviet Union was a "counterintelligence state," that is, an enterprise in which the premier function of the "organs" was to preserve the exclusive claims to power of the Communist Party and its ruling cadres. This "counterintelligence state" fixated on enemies, real and imagined, domestic and foreign. From the very first days of the USSR the intelligence services, as they were mistakenly labeled in the west, were imbued with a counterintelligence character, as was the whole of state and society. Soviet foreign intelligence had the demeanor and feel of external counterintelligence, a characteristic inherited in part from its Okhrana predecessor of Tsarist days.[20]

The Bolshevik regime was a conspiracy come to power. The Soviet Union in practice was a seventy-one year old counterintelligence operation raised to the level of a state system. The Party and the secret police operated in a conspiratorial amalgam perpetually focused on "enemies." When this

---

19   For details on Soviet/Russian intelligence and the role of deception, see: John J. Dziak, "Soviet Deception: The Organizational and Operational Tradition," in Brian D. Dailey and Patrick J. Parker (eds.), *Soviet Strategic Deception*, Lexington, MA: Lexington Books, 1987, pp. 3-20; and John J. Dziak, *Chekisty: A History of the KGB*, Lexington, MA: Lexington Books, 1988.

20   It is no accident that the Russian Foreign Intelligence Service (SVR) dates itself to 1920, when the Foreign Department of the CHEKA was established. The KGB (State Security), now known as the FSB, celebrates 1917 as its birth year, that is, when the CHEKA was established. The CHEKA, KGB, and FSB fundamentally were/are counterintelligence services. When Lenin and Dzerzhinsky decided in 1920 to create a Foreign Department of the CHEKA, the Civil War was virtually over, the CHEKA having played a key role in the Bolshevik victory.

system projected itself, either invasively or through assistance to clients, the same structure, habits, and mentality were imposed or emulated on the receiving end.

Organic to such a counterintelligence system is the widespread practice of provocations, diversion, deception, disinformation, "maskirovka" (military focused deception), penetration, and other active measures of a highly aggressive nature (hereafter collectively referred to as deception or active measures). From the first days of the Bolshevik regime these aggressive operations were conducted on a truly strategic scale, targeting Moscow's domestic and foreign enemies, the celebrated Trust (or Trest) legend being just the more visible of numerous similar major actions.[21]

An institutional mechanism, called the Disinformation Bureau, for coordinating and orchestrating active measures was established in the GPU (state security) in 1923 by a Politburo decision;[22] its successors exist to this day in several intelligence organs, the military, and other Russian state entities, albeit with various name changes. By the end of the Cold War the Party, the secret police and the military had in place a highly structured and centrally coordinating mechanism for all aspects of active measures and "maskirovka" (military deception), to include a wide range of defensive and offensive activities. This structure and associated operations carried over intact to the successor Russian Federation.

When the Soviets mounted their thrust into the Third World in the 1950s, the security organs were dominant with the KGB in the lead. The GRU took point for the Soviet military. In the Middle East the KGB and GRU worked with their intelligence and security counterparts in Egypt (they were later thrown out by Sadat), Yemen, Syria, Libya, Iraq, Algeria, and others. The "others" included various terrorist groups throughout, but not limited to, that region (e.g., the PLO, PLFP, ETA in Spain, etc.). Moscow supplemented its presence in the Middle East and elsewhere with cadres from the East European (especially East German) intelligence services, and Cuba. Some of the Russian intelligence relationships continued after the collapse of the USSR, especially

---

21 The Trust (Trest) was a combined provocation and active measure (*kombinatsiya* in Russian) – a classic strategic counterintelligence operation simultaneously targeted against domestic and foreign intelligence enemies. The operational "style" of the Trust characterized the intelligence lessons the KGB and GRU imparted to their clients and surrogates since 1917.

22 V. A. Kirpinchenko (ed.), *Ocherki Istorii Rossiyskoy Vneshney Razvedki, Tom 2, 1917 - 1933* (Sketches form the History of Russian Foreign Intelligence, Vol. 2, 1917 - 1933, Moscow: International Relations, 1996, p. 13.

the ones with Syria and Iraq. The connections with Iraq were obvious in the events leading up to Desert Storm in 1991 and the U.S. invasion of Iraq in 2003. The Russian Federation still maintains the same close intelligence links with the Syrian Ba'athist Regime of Bashar Assad that its Soviet predecessor had with his father, Hafez al-Assad.[23] Next to Iran, Damascus and its intelligence arms probably have more intimate links to Islamic terrorists, including the Iranians, than any other Muslim state. Indeed, one of the most persistent deceptions has been the carefully fostered myth that most Islamic terrorist groups are transnational, with no direct state support. Syrian links to Jihadists in Iraq as well its decades-long provision of safe haven to numerous secular and Islamist groups (including Iraqi Ba'athists today), reflects the pattern from the Soviet era when Moscow and its surrogates both perpetuated the legend of non-connectivity to their terrorist clients. The reported January 2006 visit of Iranian President Ahmadinejad to Damascus where he met with Bashar Assad and with one of the worlds most notorious and wanted Islamic terror chiefs, Imad Mugniyeh, would fit with long Syrian practice.[24]

The old Soviet intelligence connections entailed far more than mere liaison for information sharing. Training for insurgency, terrorist, and military operations occurred both in the Middle East and back in the Soviet Union and East Europe.[25] Huge quantities of weaponry and other military equipment had flooded the region over several decades. It is no accident that the signature weapon of Islamist terrorist groups today is the AK-47 assault rifle. This intelligence and military collaboration also included the whole panoply of tech-

23 From 9 March 2005 testimony of Dr. Walid Phares before Helsinki Subcommittee of U.S. CSCE, reprinted in: Dr. Walid Phares, "The Russian Syria Connection," *Front Page Magazine*, 18 March 2005, http://www.frontpagemag.com/Articles/ReadArticle.asp?ID=17389; and Ariel Cohen, "Russian Spying for Saddam Demands a Careful U.S. Response," Heritage Foundation, Webmemo # 1023, 31 March 2006 http://www.heritage.org/Research/RussiaandEurasia/wm1023.cfm.

24 *The Sunday Times*, 23 April 2006, http://www.timesonline.co.uk/printFriendly/0,,1-524-2147683-524,00.html. *Jane's Intelligence Review* is cited along with former and current U.S. officials that a terrorist intelligence summit occurred in Syria at this time in which Mugniyeh and the Iranian President participated. Mugniyeh, who is linked to the bombing of the U.S. Embassy and Marine barracks in Beirut in 1983, the torture and execution of the CIA Beirut Station Chief William Buckley in 1984, and the 1985 highjacking of a TWA jet and murder of a U.S. Navy diver passenger in 1985. He is considered one of the most dangerous and capable Islamic terrorist operatives ever and has been intimately linked with Iran and Hezbollah. The same sources concluded that Mugniyeh is charged with overseeing Iran's retaliation against the West should the U.S. attack Iran's nuclear weapons' facilities.

25 Among the premier KGB/GRU sites in the USSR for such training were Balashika in European Russia and Simferopol in the Crimea.

niques in deception, disinformation, maskirovka, etc., at which the Soviet and, later, Russian intelligence services excelled. The Iraqis demonstrated their adeptness at these lessons before and during Desert Storm (witness their success in evading U.S./Allied searches for their elusive mobile SCUD missiles) and in the interwar period in their cat and mouse games with UN inspectors and U.S. intelligence searches for their elusive weapons of mass destruction (WMDs). It must be remarked, too, that the long years of the anti-Soviet Jihad in Afghanistan provided hands-on training in which thousands of mujahideen from around the Muslim world absorbed Soviet battlefield intelligence, counterintelligence, and associated deception experience – and how to counter it. In sum, decades of Soviet intelligence warfare, technology, influence, tutelage, and presence were added to existing Islamic Jihadist traditions in the world of deception and counter deception.

As Arab nationalist and pan-Arabic dreams died in the repeated failures in the multiple Arab-Israeli wars and the collapse of their main patron – the USSR – Islamists picked up the march. The discredited secular radical movements were either overtaken by, or morphed into, the existing Muslim Brotherhood and its spin-offs. Salafists, Al-Qaeda, other Wahhabist or Sunni radical elements, and Shia radical groups rounded out the trend. While pan-Arabic dreams may have evaporated in disillusion, Moscow's intelligence legacy has left its mark, being absorbed in today's Jihadism in ways probably unforeseen by its original Soviet craftsmen.

Resurgent, radical Islam's encounter with the twentieth century's two violent totalitarian ideologies produced a troubling legacy and residue. Whereas the victorious Allies dug Nazism out root and branch in conquered Germany through a determined program of de-Nazification, core elements of Nazi ideology nevertheless have prospered in radical (and not so radical) Islamic thinking especially in its noxious anti-Semitism. Iranian President Ahmadinejad's ugly rants about Israel and the holocaust are representative of this phenomenon. The longer-lived infatuation with Soviet Communism, especially in its conspiratorial, Bolshevik elitist and active measures dimensions, added still more layers of totalitarian style to a militant theocratic mindset inclined to millenarian thinking to begin with. The fact that a process of de-Communization never materialized to exorcize seventy-one years of Soviet mass murder and other criminality,[26] seems to have had the effect of

---

26 This statement applies to the former Soviet Union. Selected former Communist East European states – Poland, the Czech Republic, the Slovak Republic, and Hungary in particular – have enacted limited measures to cleanse their political systems of the Communist legacy.

sanctioning the totalitarian legacy gifted by Moscow's long association with radical Middle Eastern states and movements. The thuggery of the Iraqi and Syrian Ba'athists, Libya's Qaddafi, the Iranian Mullahs, and the various Islamist terrorist groups is but one example of the bitter fruit of that long romance.

Stratagem in the Islamic tradition

From its earliest history Islam has practiced what Westerners label stratagem, deception, dissimulation, concealment, etc., in its dealings with not only the Infidel but with other Muslims as well.[27] Islamic scripture itself could be referenced for an early example of this with apologists' oft quoted Koranic verse, "There is no compulsion in religion" (Koran, Surah II: 256),[28] as an example of alleged Islamic tolerance for other faiths. What is not admitted or mentioned is the doctrine of abrogation in which, for all practical purposes, that particular Surah is cancelled out by harsher, more intolerant, and more violent verses coming from chronologically later revelations. Other similar examples in the deceptive use of the Koran and Hadith abound.

"Taqiyya" and "kitman" are the relevant and operative Islamic terms covering such practices and have been used for centuries; they acquired re-newed purchase with the ascendancy of Islamist terrorism in recent decades. Taqiyya – deliberate dissimulation or deception – was originally developed in Shia Islam as a defensive mechanism against Sunnis, but has since come to be accepted practice by both branches of Islam; it initially entailed masking one's true religious beliefs, especially in the face of danger by those hostile to such beliefs.[29] Closely related but lower on the ladder of deception is kitman, akin to mental reservation or, as one wag put it, holy hypocrisy. Still another and related technique in this genre and heavily used by Islamist apologists (even Bin Laden himself) is "tu quoque," a Latin phrase for a common fallacy in ar-gument and debate wherein a defense against a charge is made by turning the charge or critique back against the accuser in a manner that is irrelevant

---

None of these actions, however, equaled the scope and intensity of the de-Nazification mea-sures of an earlier era. Still, they far exceeded anything the Russians timorously undertook.

27 This part of the discussion draws from several articles from the daily offerings of Jihad Watch, and DhimmiWatch, http://www.jihadwatch.org/dhimmiwatch/; and "Taqiyya and Kitman: The Role of Deception in Islamic Terrorism," http://www.ci-ce-ct.com.main.asp.

28 Marmaduke Pickthall (commentator/translator), *The Meaning of the Glorious Koran*, NY: Alfred A. Knopf/Everyman's Library, 1930/1992, p. 59.

29 Alalmah Tabatabai, *Shi'ite Islam*, Albany, NY: State University of New York Press, 1975, p. 223.

to the truth of the original charge. It is the standard "red herring" of politics and yellow journalism but is a good diversionary tactic as well as a broader strategy since the accuser quickly becomes the accused. It works especially well against contemporary Western societies subject to fads of guilt-ridden political correctness. Islamists and their apologists use it quite effectively in the media during debates or especially in sound bite interviews, press releases, and international discourse. The more audacious the taqiyya, kitman, or tu quoque item, the more likely it will be successful. For instance when an Islamist apologist insists that Jihad is merely a spiritual striving rather than Jihad of the sword, and hides the fact that the former definition is a relatively recent one in Islam (a little over a century), he is practicing kitman.[30]

It is not only in the higher political and public realms that these Islamic practices have practical utility and psychological impact. In the operational realms of terrorism and counterterrorism Islamists have grasped their applicability: "Al-Qaeda training manuals... carry detailed instructions on the use of deception by terrorists in Western target countries.... The study of taqiyya and kitman is crucial to an understanding of Islamic fundamentalism and terrorism ranging from the issuing of false terrorist threats, operational and strategic disinformation issued by Al-Qaeda in the form of 'intelligence chatter,' to the use of taqiyya and kitman by terrorists during interrogation and the use of systematically misleading expressions concerning Islam and terrorism by Muslim spokesmen."[31]

Western interrogators of captured Islamists are at a distinct disadvantage if they have neither the language nor a foundation in Islamic history and culture to enable them to grasp the subtleties of when any of these deceptive techniques are being worked against them. For instance, it is especially dangerous to rely exclusively on indigenous translators as funnels or screeners for intelligence from such sources. Indeed, this vulnerability extends well beyond the unprepared intelligence interrogator to the higher reaches of the intelligence, counterintelligence, and policy communities where a struggle is still underway to determine who the enemy actually is, so intimidated are we about naming him. In the face of such cognitive dissonance, Islamists need not fear that we'll soon catch on to the subtleties of taqiyya, kitman, or tu quoque. It is not that Islamists are so deviously adept at their brand of strategic

---

30 Jihad Watch, 13 January 2005.

31 As quoted in: Richard H. Shultz, Jr. and Ruth Margolies Beitler, "Tactical Deception and Strategic Surprise in Al-Qaeda's Operations," *MERIA Journal*, Vol.8, No.2, June 2004, p. 4. http://www.meria.idc.ac.il/journal/2004/issue2/jv8n2a6.html.

and tactical deception. We may be exhibiting some of the same symptoms that Stalin portrayed in his refusal to recognize what was facing him across his western frontiers. German deception operations against him before June 1941 were not so sophisticated that Soviet intelligence couldn't detect them. Many capable NKVD and GRU officers did precisely that – but Stalin didn't like what they had to say because of his paranoia against his own subordinates, his deep preconceptions, his ignorance of the "other," his denial of the obvious, and his willful self-deception. Likewise, Islamic ambitions and associated stratagems have been part of the world's story for almost fourteen hundred years and have been rubbing up against the frontiers of Western Civilization for most of that time. Now due to heavy Islamic immigration and declining western demographics militant Islam is a significant presence within those frontiers – yet we still can't fathom it.

Conclusions

It has been advanced by serious students of warfare that what Islamists in general and Al-Qaeda in particular are pursuing is a new paradigm of war: Fourth Generation Warfare in which nation-states and all the associated conventions of warfare are being eclipsed by non-state players who won't acknowledge the nation-state's monopoly over armed force.[32] In such a paradigm decentralization, information warfare, pervasive deception and other time-proven counterintelligence actions, religious-based millenarian ideologies, the use of unconstrained violence recognizing no boundaries between combatants and civilians – all of these and much more characterize this new stage of warfare. This writer takes no issue with such a construct, although it does seem that the non-state dimension is somewhat overdrawn given the behind-the-scenes roles of the Syrian and Iranian intelligence services, the funding and religious play of the Wahhabist Saudi state, or the spoiling role of the Russians especially in view of their relations with Syria, Iran and, by extension, their terrorist client groups.

Rather, what is suggested is that while there very well may be Fourth Generation Warfare underway, it is a reassertion of a far older tradition: the counterintelligence imperative of traditional societies and cultures – this time in the face of a globalization destructive of old certainties. The resurgence of

---

32 See, for instance, Martin van Creveld, *The Transformation of War*, NY: Free Press, 1991; and Shultz and Beitler, ibid, pp. 2-3. The Shultz/Beitler study is an excellent analysis of how groups like Al-Qaeda have grasped and used deception and other asymmetrical advantages against the superior military and technological capabilities of the U.S.

an aggressive Islam following the failures of Arab nationalism *cum* socialism, Pan-Arabism, and the romance with the Soviet Bloc is part of this reassertion. It looks to an older Islamic patrimony based on the certainties of its theocratic faith and the lost glories of the Caliphate as the redeeming palliative to a rootless modernity. It finds comfort in Islam's own proven counterintelligence heritage of conspiracy, provocation, and deception crafted in centuries-long wars against both the Infidel and fellow Muslims. Fourteen hundred years of unchanging tradition buttressed by the hard certainties of an unmediated faith issued directly from Allah is a foundation for action, a foundation not readily reformed, modified, or dismissed. Resurgent Islam will naturally use the techniques of asymmetric struggle associated with Fourth Generation Warfare because they are precisely the techniques common to traditional societies in their confrontation with more powerful or technically advanced opponents. Deception especially is a featured tool of such societies because it is cheap, cerebral, ready-to-hand and a weapon of choice for the weaker combatant against the more powerful, but usually smug, opponent.

The residual influence of Nazi Germany and Soviet Russia on Islamism may be seen precisely in the assimilated features of the counterintelligence state absorbed by both radical Islamic movements and radical Islamic regimes: the multiplicity and redundancy of intelligence and counterintelligence services with counterintelligence being the preferred tendency; fixation with conspiracies and incessant conspiratorial intrigue; provocation and associated deception; conspiracy-laced propaganda and very sophisticated information warfare campaigns; draconian police state tactics, this time justified by theocratic strictures vice party dogma. In its drive to nuclear power status Iran, especially, has shown adeptness at deception in masking the weapons side of its program, and in information warfare and propaganda with its bombast of military prowess aimed at strong anti-war sentiment in the U.S.

In responding to a resurgent Islamism that is prosecuting a new generation of warfare especially in its deceptive dimensions, U.S. intelligence faces several difficulties, not all of them within its powers to address. With regard to the latter, public opinion, especially in its elite manifestations, is conflicted about linking a major world faith with terrorism and other violence. Coupled with fixations with political correctness and widespread nonjudgmental relativism, strong elements of U.S. opinion seem still to be in a pre-9/11 mindset. There is a palpable reluctance in many quarters to admit that a conflict along civilizational/cultural boundaries is underway.[33] A deep reluctance to

---

33   See Samuel P. Huntington, *The Clash of Civilizations and the Remaking of the World Order*,

acknowledge this is seen in the censored language of public discourse, i.e., the use of generic "terrorism" *vice* Islamists, radical Islam, etc.

What intelligence can attend to more vigorously and more cerebrally are a recognition of and response to the following:

•   A counterintelligence imperative and mentality suffuses traditional cultures from which resurgent radical Islam emerges.

•   Islamism embraces a deeply counterintelligence ethos that manifests stratagem, deception, conspiracy; Islamism is the twenty first century heir to the counterintelligence state traditions of the totalitarian systems of the last century.

•   The Islamist threat cannot easily fit the KGB paradigm, the secret police/state security model against which U. S. intelligence and counterintelligence have been defending since World War II. Countering the current threat is not just a matter of frustrating the case officer and his recruit(s).

•   There is a lack of a serious and sophisticated counterintelligence tradition in the U.S. that degrades our coming to grips with the subtleties of Fourth Generation Warfare. Counterintelligence, deception, and counterdeception are among the most cerebral elements of the intelligence craft, yet remain the orphans of the trade. Recent official commissions, especially the WMD Commission in its report issued last year, seem to be saying pretty much the same thing.

Sophisticated counterintelligence is key to both counter deception and counterterrorism. Only then will a revivified U.S. intelligence culture be able to grapple with a religiously rooted war in which deception is organic, even to the enemy's portrayal of itself. The WMD Commission had it about right.

NY: Simon and Schuster, 1998. When Huntington first proposed in 1992, the notion that the world would divide along civilizational vice ideological boundaries, he was roundly denounced. Since 9/11 the polemics have abated.

18

Fourth Generation Warfare Evolves, Fifth Emerges

Col. T. X. Hammes, USMC Ret.

Seventeen years ago, a small group of authors introduced the concept of "Four Generations of War." Frankly, the concept did not get much traction for the first dozen years. Then came 9/11. Some of the fourth-generation warfare (4GW) proponents claimed that the Al-Qaeda attacks were a fulfillment of what they had predicted. However, most military thinkers, for a variety of reasons, continued to dismiss the 4GW concept. In fact, about the only place 4GW was carefully discussed was on an Al-Qaeda website. In January 2002, one 'Ubed al-Qurashi quoted extensively from two Marine Corps Gazette articles about 4GW.'[1] He then stated, "The fourth generation of wars [has] already taken place and revealed the superiority of the theoretically weak side. In many instances, these wars have resulted in the defeat of ethnic states [duwal qawmiyah] at the hands of ethnic groups with no states."

Essentially, one of Al-Qaeda's leading strategists stated categorically that the group was using 4GW against the United States – and expected to win. Even this did not stimulate extensive discussion in the West, where the 9/11 attacks were seen as an anomaly, and the apparent rapid victories in Afghanistan and Iraq appeared to vindicate the Pentagon's vision of high-technology warfare. It was not until the Afghan and Iraqi insurgencies began growing and the continuing campaign against Al-Qaeda faltered that serious discussion of 4GW commenced in the United States.

Yet today, even within the small community of writers exploring 4GW, there remains a range of opinions on how to define the concept and what its implications are. This is a healthy process and essential to the development

---

1 See William S. Lind, et al, "The Changing Face of War Into the Fourth Generation," *Marine Corps Gazette*, October 1989. See also Thomas X. Hammes, "The Evolution of War: The Fourth Generation," *Marine Corps Gazette*, September 1994.

of a sound concept because 4GW, like all previous forms of war, continues to evolve even as discussions continue. That brings me to the purpose of this article: to widen the discussion on what forms 4GW may take and to offer a possible model for the next generation of war: 5GW.

Developments in 4GW

Current events suggest that there are a number of ongoing major developments in 4GW: a strategic shift, an organizational shift, and a shift in type of participants.

Strategic shift. Strategically, insurgent campaigns have shifted from military campaigns supported by information operations to strategic communications campaigns supported by guerrilla and terrorist operations. While there is no generally agreed upon definition of 4GW, according to the definition I wrote in 2003, "Fourth generation warfare uses all available networks – political, economic, social, and military – to convince the enemy's political decision makers that their strategic goals are either unachievable or too costly for the perceived benefit. It is an evolved form of insurgency." The key concept in this definition is that 4GW opponents will attempt to attack the minds of enemy decision makers directly. The only medium that can change a person's mind is information. Therefore, information is the key element of any 4GW strategy. Effective insurgents build their plans around a strategic communications campaign designed to shift their enemy's view of the world.[2]

It is clear that many insurgent groups understand this fact. Hezbollah's strategy during the 2006 summer war with Israel is an excellent example. During the fighting, they focused not on damaging Israel, but on ensuring they were perceived as defying the most powerful army in the Middle East. Thus, the fact that Hezbollah fired as many rockets on the last day of the war as the first was critically important. They know 122mm rockets are notoriously inaccurate and cause little damage, but the rockets are highly visible. Their appearance "proved" the powerful Israeli Air Force and Army had not hurt Hezbollah badly.

---

2 I have intentionally chosen to use "strategic communications campaign" instead of "information campaign" for two reasons. First, the Pentagon's definition of information operations states that "the principal goal is to achieve and maintain information superiority for the U.S. and its allies." Unfortunately, it sees information primarily as computer and communications security and exploitation. Second, the very phrase "information operations" leads one to focus on the tactical or operational level. In contrast, "strategic communications" by definition falls at the strategic level of war and subsequent operational and tactical efforts must support that strategic approach.

Once the fighting stopped, Hezbollah showed an even greater grasp of strategic communications. While the West was convening conferences to make promises about aid at some future time, Hezbollah representatives hit the streets with cash money and physical assistance. To the Arab world, the contrast could not have been clearer. When Israel needed more weapons, the United States rushed them in by the planeload. When Arab families needed shelter and food, we scheduled a conference for some future date. Hezbollah acted – and gained enormous prestige by doing so. To ensure they continued to dominate this critical communications campaign, Hezbollah physically prevented other agencies from distributing aid in Hezbollah areas. The message was clear – Hezbollah was sovereign in its territory and focused on its people. The contrast between that message and the usual apathy of Arab governments to their people's needs was stunning.

Hezbollah is not an isolated case. The high quality and enormous variety of insurgent web sites indicate many, if not most, insurgent groups understand the imperative of executing an effective strategic communications campaign when trying to drive out an outside power. In contrast, the United States continues to flounder in its efforts at strategic communications.

This shift from Mao's three-phased insurgency to a strategic communications campaign has been developing since Ho Chi Minh's successful effort at breaking America's political will over Vietnam. Today, it is clearly the primary choice of insurgents faced with outside powers. However, just as Mao's strategic concept included a Phase III conventional battle to defeat the government, the new "coalitions of the willing" know they will also face a final phase. Theirs will be the civil war to decide who among them will control the country after the outside power is gone. Unfortunately, post-Soviet Afghanistan and today's Gaza Strip show that once the outside power is driven out, the civil war quickly devolves from 4GW to a traditional 2GW war of attrition.

Organizational shift. The emergence of civil war as a part of insurgency is based on the major organizational shift that has occurred since Mao formulated his concept. It reflects the continuous, worldwide shift from hierarchical to networked organizations. While the Chinese and Vietnamese insurgencies were hierarchies that reflected both the social organizations of those societies and the dominant business and military organizations of the time, recent insurgencies have been networked coalitions of the willing. For instance, in Iraq, there is no unifying concept among the various insurgent groups except to get Americans out of the country. While some of the more centrist groups could form a coalition government, clearly the Sunni Salafists and Shia religious militias cannot coexist if we are driven out; in fact, they are already fighting a civil war in anticipation of our departure. Other groups, such as criminal networks, cannot tolerate a strong central government of any kind – unless it is thoroughly corrupt and lets them continue their criminal activities.

The rise of networked coalitions is in keeping with the fact that both the societies in conflict and the dominant business organizations of our time are networks. Like society as a whole, insurgencies have become networked, transnational, and even trans-dimensional. Going beyond simple real-world networks, some elements of their organizations exist in the real world, some in cyberspace, and some in both dimensions.

Shift in participants. As part of the organizational shift, we have seen a change in who is fighting and why. It is essential for us to understand that, even within a single country, the highly diverse armed groups that make up a modern insurgency have widely differing motivations. Studying the motivation of a group gives us a strong indication of how that group will fight and what limits, if any, it will impose on its use of force. The UN's Manual for Humanitarian Negotiations with Armed Groups states, "In terms of founding motivations, armed groups generally fall into three categories: they can be re-actionary (reacting to some situation or something that members of the groups experienced or with which they identify); they can be opportunistic, meaning that they seized on a political or economic opportunity to enhance their own power or positions; or they are founded to further ideological objectives."[3]

Reactionary groups often form when communities feel threatened. They tend to be sub-national or national groups that operate in specific geographic areas and attempt to protect the people of those areas. In essence, these armed groups represent a return to earlier security arrangements; they are the result of a state's failure to fulfill its basic social contract of providing security for its population. The ethnic-sectarian militias we have seen develop around the world in response to insecurity are reactionary groups. The Tamil Tigers and Badr Militia are typical of the type.

Reactionary groups need to protect populations but lack the military power to do so. As a result, they usually resort to 4GW – but generally use only conventional arms. While highly effective, such weapons are familiar to Western armies and thus easier to anticipate and defeat. Reactionary groups also tend not to be a threat outside their areas since they are focused mainly on

---

3 United Nations Manual for Humanitarian Negotiations with Armed Groups, 16.

defending their own people. However, they still conduct sophisticated communication campaigns to defeat outside powers.

Opportunistic groups spring up to take advantage of a vacuum to seize power or wealth. Criminal by nature, these groups have been around for centuries. What is different now is that commercially available weapons allow them to overmatch all but the most well-armed police – they are even a match for the armed forces of some nations. Opportunistic groups include organizations like Mara Salvatrucha 13 (MS-13) and, increasingly, the Irish Republican Army (IRA). Opportunistic groups conduct their own strategic communications campaigns, usually citing a religious or national cause to claim legitimacy for their criminal activities.

A third great motivator, ideology, gives birth to the most dangerous armed groups – organizations like Al-Qaeda, Aryan Brotherhood, and Aum Shinrikyo. Ideological groups are more dangerous to the United States than reactionary or opportunistic groups because of their no-limits approach to conflict. In the past, they have used society's assets against it. From Timothy McVeigh's bomb made of fertilizer and diesel fuel to Al-Qaeda's employment of airliners, ideological groups tend to be highly creative in their attacks. They are more likely to use society's infrastructure – chemical plants, mass shipments of fertilizer, even biotechnology – as weapons of mass destruction than groups motivated by self-defense or opportunism.

Of even more concern is the fact that ideological groups are essentially impossible to deter. First, their "cause" provides moral justification, and sometimes a moral requirement, to use any available weapon. Second, they have no return address, so they do not fear massive retaliation—if Al-Qaeda detonates a nuclear device on U.S. soil, where exactly do we fire our nukes in return?

Ideological groups will not be deterred even by the danger inherent in the use of biological weapons. While other groups may hesitate to release a contagious biological agent for fear of killing their own people, ideological groups believe the higher power guiding their actions will either protect their members or call them home for their earned reward. Thus, the combination of extraordinarily rapid advances in biotechnology and the spread of ideologically driven armed groups represents a major threat to the global population.

While the UN manual cites three kinds of differently motivated insurgent groups, recent developments point to the advent of a fourth: a hybrid spurred by a blend of reactionary, ideological, and/or opportunistic motivations. Sometimes these groups are reactionary or ideological, but then turn to crime for funding. Al-Qaeda, for instance, is primarily an ideological group that has become increasingly opportunistic in order to subsidize its operations. The IRA started as a reactionary group, but it too has increasingly turned to crime – and may actually have moved from a reactionary to a purely opportunistic motivation.

Another kind of hybrid is the ideological group that finds itself de facto ruler of an area: by taking charge, it becomes bound to protect the community, just as reactionary groups must. The Jaysh Al Mahdi militia in Iraq is one such example.

Some groups can even fall into all three categories. For instance, Hamas and Hezbollah provide protection, espouse an ideology, and participate in crime for funding. In fact, most armed groups now use crime to fund operations.

The sad truth is that there is a truly alarming variety of armed groups active in the world today. Understanding their motivations, methods, and goals is becoming increasingly difficult.

Weapons of mass destruction

Iraq has seen the development of another major refinement of 4GW: using more or less basic materials to create weapons of mass destruction (WMD). While Western intelligence agencies have long worried that the Iraqi insurgents would use industrial chemicals, only recently have they used chlorine as part of their attacks. Much like World War I's combatants, the insurgents had to learn that it takes the right conditions and huge quantities of gas to create large numbers of casualties; however, they and their brethren around the world have shown a distinct ability to learn from each other, and now the Iraqi attacks are becoming increasingly effective. Although it might be nearly impossible to repeat, Al-Qaeda's 9/11 operation with airliners was certainly a massively destructive attack forged from unconventional (nonnuclear, non-chemical, non-biological) WMD materials. In contrast to 9/11, the extensive availability of toxic industrial chemicals means that massive chemical attacks can be duplicated in many areas of the world.

What makes this WMD-like development particularly troubling is that some terrorist websites have discussed using chemical plants or shipments to cause the numbers of casualties that occurred, for instance, in Bhopal, India, in 1984, when fumes from an industrial gas leak enveloped the city, killing thousands. The 1947 disaster in Texas City, Texas, when a ship with 8,500 tons of ammonium nitrate on board blew up in port and killed nearly 600, is another possible template for achieving WMD-like effects. If either incident had been intentional, it would have qualified as a WMD attack. This move toward unconventional WMD development, coupled with the trend shown by the Iraqi insurgents' increasingly effective use of chlorine, presents an immediate and major danger to U.S. interests both at home and overseas.

Another new player: private military companies

A largely overlooked development in warfare is the exceptional increase in the use of private military companies (PMCs). These organizations have always been around, but during the last two decades they have become central to the way the United States wages war. There has been very little consideration given to how PMCs might impact international relations in general and war in particular. While we have focused on the monetary and political cost-cutting benefits of PMCs, other nations are discovering creative ways to use them to avoid normal international constraints on the use of force.

Of particular concern is the use of armed contractors. The length of this article prevents a full exploration of the numerous implications that flow from the increased use of armed contractors, so I will simply offer some thoughts to start a discussion. For instance, How does one hold a country accountable for the actions of an armed PMC? How will these companies change the face of armed conflict? What impact will they have on the relationship between the rulers of resource-rich countries and their populations? Can they be employed to provide bases or major forward-deployed combat assets?

PMC spokesmen have continually reassured us that their companies are responsible organizations that are working with governments to devise effective regulations for PMC employment. This is, in fact, true. However, while the United States has moved to increase the accountability of such companies through regulations and contracts, these methods have yet to be seriously tested. Further, much like the shipping industry avoids regulation by registering under flags of convenience, we can expect PMCs to do the same: if regulations interfere with how they wish to operate, they will move to another country or even dissolve their corporations and start again as different legal entities in different countries. We have already seen a number of PMCs do exactly that.

The sudden presence of PMCs in numerous conflicts worldwide presents some interesting challenges to the international community. In the more than 300 years since the Treaty of Westphalia, we have developed diplomatic, economic, and military techniques for dealing with crises created when nation-states use armed force – or even threaten to use it. We do not have such mechanisms in place when nation-states or even private individuals employ armed contractors. If China had announced that it planned to send multiple field armies to Angola to assist with security and construction there, the UN would at least have opened up a dialogue. Yet a Chinese company has signed a contract to do just that, except that it will substitute 850,000 armed and unarmed contractors for the field armies. This event has simply not shown up in international discussion. It is particularly interesting because China has just signed a 10-year contract with Angola to purchase oil at $60 a barrel. While the contractors are not an official branch of the Chinese Government, their presence clearly puts China in position to "resolve" any disputes with the Angolan Government over that contract. Thus, thanks to the creative use of PMCs, brokering agreements between nation-states and even the process of intervening to resolve disputes between parties has moved outside the international system. How does the UN respond to a contract dispute between an armed private company and a government?

Another interesting development is that "governments" of countries with resource-rich areas can employ PMCs to seize and hold the rich areas while they ignore the rest of the country. We have already seen this with local militias and "blood diamonds," but have not seen it applied in a systematic way. That may be happening now in the Sudan, where the Sudanese Government has hired Chinese firms to secure Sudan's oil facilities. These firms not only provide reliable security, but also have no qualms about how the Sudanese Government chooses to conduct its internal affairs. By using PMCs, a very small minority can control a country without any regard to the needs of the majority. A clique can always seize power through a coup, but it takes trusted security forces to keep the resulting governments in power. In some parts of the world, security forces are likely to be loyal to their own clans or tribes, so the government must take care of those tribes. Now, though, governments have the option of hiring an effective PMC and completely ignoring any parts of the country that are not profitable – they won't need the people to ensure their continued rule. The result will be a significant increase in the ungoverned and desperately poor areas of the world. Also reinforcing the power of an oppressive minority is the international community's policy of dealing with whatever gang controls a country's capital city. With little likelihood of outside intervention, the oppressed and poor will have to resort to violence.

PMCs can also be used to establish forward operating bases or can even be deployed as forward forces. In the same way the British used the East India Company to establish a navy, an army, and supporting bases in India, other nations such as China are using commercial entities throughout the world to protect or advance their interests. Chinese PMCs already constitute a major ground presence in Africa, and with Chinese commercial entities building ports all along the shipping lanes from the Middle East to China, China could employ naval PMCs, at least nominally, to provide security against pirates. In fact, in early March the Chinese signed a contract with Somalia to train and equip a Somali coast guard. Such naval forces will obviously need maintenance and support facilities, which the companies will build. In effect, China's PMCs can establish a chain of naval facilities complete with ships near the chokepoints of major sea routes. PMCs cannot be easily categorized as belonging to a particular generation of war. Rather, they are a tool that can be used in a wide variety of ways. But because 4GW succeeds by avoiding an opponent's military strength, PMCs offer the intriguing possibility of a weak country employing them in a 4GW manner, so that war doesn't look like war, but like business.

The final alarming fact about PMCs is that they are businesses. As such, they compete by focusing on quality, reliability, and cost. China can match Western firms on the first two and, based on a huge population of unemployed young men, can severely undercut Western firms on cost. Further, China has a huge incentive to subsidize businesses like PMCs: its one child policy has resulted in over 20 million more Chinese men of marriageable age than Chinese women.

Criminals are yet another player in 4GW. Most 4GW discussions still focus on politically motivated insurgent groups, however, as discussed in the 1989 *Gazette* article on 4GW, criminal organizations are using 4GW techniques. A good example is Mara Salvatrucha 13 (MS-13). This organization started out primarily as a criminal movement, but it is now establishing effective political control in widely scattered locations. From some communities in El Salvador and Honduras to neighborhoods in American cities and even some American suburbs, MS-13 is creating sovereignty in non-contiguous territory. Much like their commercial predecessors the Hanseatic League, MS-13 has used violence and wealth generated by trade (primarily drugs) to create enclaves within national territories.

State use of G4W

China's employment of PMCs is a clear example of a state using 4GW. Iran has taken a very different approach. Last summer, it introduced the West to the concept of lateral asymmetric escalation. As the United States continued to raise the pressure for UN action in response to Iran's nuclear program, Iran seized the opportunity presented by the Israel-Hezbollah confrontation in Lebanon to change the discussion. While we do not think that Iran instigated the war, we know it has considerable influence over Hezbollah and certainly provided extensive support to that group's efforts against the Israelis. In Hezbollah, Israel faced a 4GW enemy that made effective use of relatively high-technology weapons to challenge Israel's assumed military superiority. External to Lebanon, Iran cooperated with Syria to provide extensive logistical and perhaps intelligence support to the Hezbollah command. Because the United States and UN apparently can deal with only one crisis at a time, Iran was able to use the conflict in Lebanon in a 4GW manner to stop action against its nuclear program. Obviously, this was not a long-term solution for the Iranians, but it furthered their apparent strategic goal of buying time to develop a nuclear weapon.

4GW updated

Since the 1989 *Gazette* article, the Afghan and Iraqi insurgents have continued to shift their strategic focus to the 4GW aspect of strategic communications. Organizationally, the insurgents are evolving into an ever-increasing variety of armed groups linked into coalitions of the willing. Also, the types of players and their motivations have changed significantly over time.

As a result, the coalitions of the willing we are facing in Iraq and Afghanistan are much more challenging than their monolithic predecessors. The proliferation of motivations and merging of ideological, reactionary, and opportunistic groups makes it increasingly difficult to tell who is fighting and why. Fortunately, the bottom line remains effective security and governance for the people, and the new counterinsurgency field manual (FM 3-24, Counterinsurgency) provides solid guidance on how to achieve that. Unfortunately, the sheer number of people involved in the two conflicts precludes the United States from realizing the recommended ratio of one security officer to every 50 citizens that has generally meant success in the past. To deal with the numerous changes in 4GW, we will have to find new ways to provide security while building the political coalitions that are the only way to defeat an insurgency. We will also have to apply our diplomatic, economic, and political resources more broadly and effectively than we have done in the past to deal with the expanding nation-state use of 4GW.

Fifth generation warefare

Military institutions and the manner in which they employ violence depended on the economic, social, and political conditions of their respective states.
—Clausewitz[4]

Like always, the old generations of war continue to exist even as new ones evolve. Today, we see grim 2GW firepower-attrition battles in parts of Africa even as the first hints of 5GW emerge. This should not be surprising — countries that lack the political, social, and economic systems to support new forms of war will continue to use the older forms. Yet a new generation must also evolve and, given the fact that 4GW has been the dominant form of war-fare for over 50 years, it's time for 5GW to make an appearance. We should be able to get some idea of what this new form of war will be by examining how political, social, and economic systems have changed since 4GW became dominant.

Politically, there have been major changes in who fights wars. The trend has been and continues to be downward from nation-states using huge, uniformed armies to small groups of like-minded people with no formal or-ganization who simply choose to fight. We have slid so far away from national armies that often it is impossible to tell 4GW fighters from simple criminal elements. Many of the former are, in fact, criminal elements — either they use crime to support their cause or they use their cause to legitimize their crime.

Economically, we have seen a steady increase in the power of infor-mation. Insurgent groups have seized on the improving information grid to execute the strategic communications campaigns that are central to their vic-tories. The content and delivery of information has accordingly shifted from the mass propaganda of Mao to highly tailored campaigns enabled by the new methods of communication and new social patterns. Insurgents have been quick to exploit such powerful communication tools as the cell phone and the Internet for recruiting, training, communicating, educating, and controlling new members. They have shifted from mass mobilization to targeted indi-vidual mobilization.

---

4 Carl von Clausewitz, *On War*, ed. Michael Howard and Peter Paret (Princeton, NJ: Princeton University Press, 1989), 6.

Today's key businesses are becoming ever more productive because of their access to or manipulation of information. One result has been a proliferation of small companies that have created great wealth, a phenomenon in accordance with the long-term trend of power devolving downward to smaller entities—whether they are business or military. The epitome of this tendency is that just two guys essentially created Google.

Communications is not the only burgeoning sector with implications for 5GW. Two industries with even greater potential to change our world – biotechnology and nanotechnology – are on the verge of huge growth. In many ways military and business problems are merging as the world becomes more interconnected and power is driven downward. In 2006, a group of about 20 angry Nigerians took hostages from a Shell oil platform in the Gulf of Guinea. Shell shut down its Nigerian Delta production and world oil prices rose dramatically. The interconnected world is highly vulnerable to disruptions in key commodities, and business issues can very rapidly become matters of serious international security. This is not the same as in the old banana wars, when Marines were consistently committed to protect American interests that mattered only to a few stockholders. Today, very small armed groups can impact the entire world's economy immediately and dramatically.

Socially, we have seen a major shift in how communities are formed. People are changing allegiance from nations to causes, a trend dramatically accelerated by Internet connectivity. In fact, many people are much more engaged in their online causes than in their real-world communities. Of particular concern are members of groups who are willing to go to extremes to advance their causes – from the woman who lived in a redwood for two years to suicide bombers. Such actors place their causes above any rational analysis of the impact of their actions – and they can be found through the Internet.

In sum, political, economic, and social trends point to the emergence of super-empowered individuals or small groups bound together by love for a cause rather than a nation. Employing emerging technology, they are able to generate destructive power that used to require the resources of a nation-state.

All of these new developments are of particular concern because emerging political, business, and social structures have consistently been more successful employing nascent technology than older, established organizations. Today, two emerging technologies, nanotechnology and biotechnology, have the power to alter our world, and warfare, even more fundamentally than information technology. Most writers agree it will be 20 years or more before nanotech hits full stride, so I will not discuss it further. In contrast, today's biotechnology can give small groups the kind of destructive power previously limited to superpowers.[5]

The October 2001 anthrax attack on Capitol Hill may have been the first 5GW attack. Given the enormous investigative effort expended on finding the perpetrator(s) and the fact that we have not made a single arrest, one has to believe the attack was executed by an individual or a very small group. Had more people had been involved, someone would have leaked information or been found.

If this is a valid assumption, then we had a super-empowered individual or small group attack the legislative body of a nation-state using an advanced biological weapon in support of an unknown cause. This individual or group disrupted the operation of Congress for several months, created hundreds of millions of dollars in clean-up costs, and imposed mail screening requirements (and associated costs) that are still in effect today – not a bad payoff for a few ounces of anthrax and some postage.

The anthrax attack provided stark evidence that today a single individual can attack a nation-state. Over time, the combination of political motivation, social organization, and economic development has given greater and greater destructive capability to smaller and smaller groups. While some technologists thought we had reached a peak of destructive power with the advent of thermonuclear weapons, the fact remains that creating and delivering such weapons required an elaborate and expensive developmental effort. By contrast, the following recent developments suggest that the potentially massive destructive power of bio-weapons is within reach of motivated groups:

•   Three years ago, a team led by Dr. Craig Venter created a functioning virus from off-the-shelf chemicals. Venter's team selected a specific virus, purchased the necessary genetic base pairs to make the virus, and then "assembled" the pairs into a functioning synthetic virus. All of the materials and equipment the team used are commercially available without restrictions. Venter has predicted that what took an elite team and a very well-equipped lab to do the first time could be done by any competent graduate student in a university lab in less than a decade.

•   Paul Boutin, a science writer, decided to take up Venter's "challenge."

---

5 There has already been extensive discussion about cyber attacks, so I will not deal with that threat in this short article. However, such attacks are an option for small groups-to include the physical destruction of key fiber-optic switches and cables using simple breaking-and entering-techniques.

Despite not having been in a biology lab since high school, Boutin, with a little guidance from Dr. Roger Brent to keep him out of dangerous experiments, created glowing yeast. While yeast is not smallpox, the equipment, techniques, and nucleotides Boutin used are similar to those needed to create smallpox from its base pairs.[6]

• The complete smallpox genome has been published online and is widely available. Boutin found it in about 15 minutes.

• The nucleotides to make smallpox can be purchased from a variety of suppliers without identity verification.

• Smallpox has about 200,000 base pairs. DNA with up to 300,000 base pairs has already been successfully synthesized.

• An Australian research team heated up mousepox virus by activating a single gene. The modification increased its lethality from 30 percent to over 80 percent. It is even lethal to 60 percent of an immunized population. They posted their result on the Internet. It turns out smallpox has the same gene.

• The cost of creating a virus is dropping exponentially. If Carson's Curve continues to hold true, the cost of a base pair will drop to between 1 and 10 cents within the decade. Thus, a researcher could order all the necessary base pairs to create a smallpox virus for between $2,000 and $20,000.[7] The equipment he needs to assemble the virus will cost an additional $10,000.

• Bio-hackers are following in the footsteps of their info-hacker predecessors. They are setting up labs in their garages and creating products. Last year, a young British researcher invested $50K in equipment and produced two new biological products. He then sold his company, Agribiotics, for $22 million. We can assume hundreds, if not thousands, of young biology students are now in their basements attempting to make new biological products.

These discrete but related events mean that it is becoming increasingly easier for a small group and perhaps even an individual to create a virus such as smallpox and use it as a weapon.

Some experts have reassured us that even if a small group can create a biological virus, it is the testing, storage, and dissemination that are the most difficult steps in weaponizing a biological entity. They are right – if the creator uses traditional methods. However, a person can avoid the requirement for testing by selecting a known lethal agent, such as smallpox. He already knows it can thrive outside the laboratory. Storage and dissemination problems can

---

6 "Biowar for Dummies," http://paulboutin.weblogger.com/stories/storyReader$1439.

7 Robert Carlson, "The Pace and Proliferation of Biological Technologies," *Biosecurity and Bioterrorism: BioDefense Strategy, Practice and Science*, Volume 1, Issue 3, 2003.

be solved by tapping into the increasing trend of suicide attacks worldwide – he simply injects the smallpox directly into suicide volunteers, who become both the storage and the dissemination systems.

Using a few volunteers and commercial airlines, a terrorist group can create a near-simultaneous worldwide outbreak of smallpox. Dark Winter, an exercise conducted in 2001, simulated a smallpox attack on three U.S. cities. In a period of 13 days, smallpox spread to 25 states and 15 countries in several epidemiological waves, after which one-third of the hundreds of thousands of Americans who contracted the disease died. It was estimated that a fourth generation of the disease would leave 3 million infected and 1 million dead. The exercise was terminated at that time.[8]

It is essential to remember that not only will smallpox cause an exceptional number of deaths, but it will also shut down world trade until the epidemic is controlled or burns itself out. Given that the 2002 West Coast longshoremen's strike cost the U.S. economy $1 billion per day, the cost of a complete shutdown of all transportation will be catastrophic.

Biological weapons have the capability to kill many more people than a nuclear attack. Further, unlike nuclear weapons, which are both difficult and relatively expensive to build, smallpox will soon be both inexpensive to produce and difficult to detect until released. While I selected smallpox for this brief paper, a biologist can obviously select any of the known effective contagions. He can also attempt to create an entirely new disease. But of course no one can predict how a lab-raised disease will fare against the natural enemies it will face when released into the environment. Thus, a terrorist is more likely to use an existing disease or modify one to be more lethal. He can also release both versions of the disease – the naturally occurring virus and the enhanced virus – to insure success.

Summary

Drawing on changes in the political, economic, social, and technical fields, 1GW culminated in the massed-manpower armies of the Napoleonic era. In the same way, 2GW used the evolution to an industrial society to make firepower the dominant form of war. Next, 3GW took advantage of the political, economic, and social shifts from an industrial to a mechanical era to make mechanized warfare dominant. Fourth-generation warfare uses all the

---

8 Mark Mientka, "Dark Winter Teaches Bio Lessons," www.usmedicine.com/article.cfm?articleID=322&issueID=33.

shifts from a mechanical to an information/electronic society to maximize the power of insurgency. It continues to evolve along with our society as a whole, thus making 4GW increasingly dangerous and difficult for Western nations to deal with.

Fifth-generation warfare will result from the continued shift of polit- ical and social loyalties to causes rather than nations. It will be marked by the increasing power of smaller and smaller entities and the explosion of biotech- nology. 5GW will truly be a nets-and-jets war: networks will distribute the key information, provide a source for the necessary equipment and material, and constitute a field from which to recruit volunteers; the jets will provide for worldwide, inexpensive, effective dissemination of the weapons.

The contagion scenario I described above is among the more devastating possible, but smallpox is only one weapon a super-empowered small group could use to attack society. They may use any number of evolving technologies. The key fact to remember is that changes in the political, economic, social, and technical spheres are making it possible for a small group bound together by a cause to use new technologies to challenge nation-states. We cannot roll back those changes, nor can we prevent the evolution of war. Clearly, we as a nation, and particularly our military, are not ready to counter the coming attacks. It's time to start thinking about how we might deal with this next step in warfare.

19

Wars of Ideas and the War of Ideas

Antulio J. Echevarria II

Despite widespread emphasis on the importance of winning the war of ideas in recent strategic literature, we find few analytical studies of wars of ideas as such. With that in mind, this monograph offers a brief examination of four common types of wars of ideas, and uses that as a basis for analyzing how the United States and its allies and strategic partners might proceed in the current war of ideas.

Scoping the Problem

Simply put, a *war of ideas* is a clash of visions, concepts, and images, and – especially – the interpretation of them. They are, indeed, genuine wars, even though the physical violence might be minimal, because they serve a political, socio-cultural, or economic purpose, and they involve hostile intentions or hostile acts. Wars of ideas can assume many forms, but they tend to fall into four general categories (though these are not necessarily exhaustive): (a) intellectual debates, (b) ideological wars, (c) wars over religious dogma, and (d) advertising campaigns. All of them are essentially about power and influence, just as with wars over territory and material resources, and their stakes, can run very high indeed.

Common Wars of Ideas

Intellectual debates are disputes in which opposing sides advance their arguments, support them with evidence, and endeavor to refute the reasoning and conclusions of the other. Examples include the ongoing debate between pro-choice and pro-life advocates, and the recent dispute between the theories of "intelligent design" and evolution.

Ideological wars are a clash of broad visions usually organized around a doctrine, whether secular or nonsecular. The most popular example of an ideological conflict is the Cold War, which involved political, economic, and military competition between the United States and the Soviet Union and their respective allies.

Disputes over religious dogma are a form of intellectual debate, but they center on conflicting interpretations of sacred tenets or texts, the access to which can be, and often is, deliberately restricted or otherwise limited. Examples include the Sunni-Shi'ite split within Islam and Catholicism's East-West schism.

Advertising campaigns are contests between competing producers or vendors for "market share." The objective of such campaigns is to persuade audiences to take desired actions, such as voting for a particular candidate, visiting a certain place, or buying a specific product. A classic example is the "Cola Wars" between Coca-Cola and Pepsi-Cola.

Wars of ideas: some conclusions

Inconclusive outcomes are not unusual in wars of ideas. Opposing sides seldom change their positions based on the introduction of new evidence, or new ways of evaluating existing evidence. Thus, wars of ideas are rarely settled on the merits of the ideas themselves. Instead, they tend to drag on, unless an event occurs that causes the belligerents to focus their attention elsewhere. When conclusive outcomes do occur, they tend to follow the physical elimination or marginalization of one side's key proponents. In other cases, a major event, such as the collapse of the Soviet Union, might occur that renders one side incapable of continuing the conflict or campaign. Thus, physical events, whether designed or incidental, are in some respects more important to the course and outcome of a war of ideas than the ideas themselves.

"The war of ideas"

Diverging Approaches? Two diverging schools of thought exist on how the United States and its partners should approach the current "war of ideas" with Al-Qaeda and similar groups. The first treats the conflict as a matter for public diplomacy, defined as the "conveyance of information across a broad spectrum to include cultural affairs and political action." Accordingly, this view calls for revitalizing the U.S. Department of State, and reestablishing many of the traditional tools of statecraft. The second advocates waging the war of ideas as a "real war," wherein the objective is to destroy the influence and credibility of the opposing ideology, and neutralize its chief proponents. It calls for continuing the transformation of the U.S. Department of Defense so that it can better leverage information-age weapons.

Although each approach has merits, neither is informed by an understanding of wars of ideas as such. U.S. strategy for the war of ideas requires a more precise goal than just improving America's image. Winning a popularity contest is far less important than undermining Al-Qaeda's ability to recruit. The two aims are certainly related, but eminently separable. Success in the former does not necessarily equate to success in the latter; conflating the two aims only creates confusion.

Recommendations

• U.S. strategy for the war of ideas must be more alert to the opportunities and pitfalls introduced by physical events. For instance, the successful stabilization of Afghanistan and Iraq would have an extremely positive effect on the war of ideas, undercutting Al-Qaeda's general information campaign.

• Neither the Department of State's approach nor that of the Department of Defense should be subordinated to the other. Rather, the United States should pursue both approaches in parallel.

• Both Departments should sponsor studies and conferences that will explore wars of ideas in more depth, thereby promoting greater understanding.

• The Joint community should revise its doctrine concerning information operations, to include psychological operations and military deception. The basic assumption underpinning current doctrine is that information operations are a subset of support to military operations. Yet, in some cases, military operations might need to support information operations.

• U.S. doctrine on information operations must also acknowledge that the "information environment" is neither neutral nor static. Disparate cultural and social influences almost always ensure that diverse audiences will interpret the same information differently.

• The U.S. Army's new Human Terrain System, which helps enhance cultural awareness, is an important step in the right direction and should be supported. By developing an understanding of wars of ideas as a mode of conflict, we can fight the current battle of ideas more effectively, while at the same time better prepare ourselves to wage future ones.

www.ingramcontent.com/pod-product-compliance
Lightning Source LLC
Chambersburg PA
CBHW062158270326
41930CB00009B/1577

* 9 7 8 0 6 1 5 5 1 9 3 9 5 *